AF477586

UNITY OF SCIENCE

SYNTHESE LIBRARY

MONOGRAPHS ON EPISTEMOLOGY,

LOGIC, METHODOLOGY, PHILOSOPHY OF SCIENCE,

SOCIOLOGY OF SCIENCE AND OF KNOWLEDGE,

AND ON THE MATHEMATICAL METHODS OF

SOCIAL AND BEHAVIORAL SCIENCES

Managing Editor:

JAAKKO HINTIKKA, *Academy of Finland and Stanford University*

Editors:

ROBERT S. COHEN, *Boston University*

DONALD DAVIDSON, *University of Chicago*

GABRIËL NUCHELMANS, *University of Leyden*

WESLEY C. SALMON, *University of Arizona*

VOLUME 109

ROBERT L. CAUSEY

UNITY OF SCIENCE

D. REIDEL PUBLISHING COMPANY

DORDRECHT-HOLLAND/BOSTON-U.S.A.

Library of Congress Cataloging in Publication Data

Causey, Robert L
    Unity of science.

    (Synthese library; 109)
    Bibliography: p.
    Includes indexes.
    1. Science—Philosophy. 2. Identity. 3. Knowledge,
Theory of. 4. Theory (Philosophy)
I. Title.
B67.C33   501   76-57970
ISBN 90-277-0779-0

Published by D. Reidel Publishing Company,
P.O. Box 17, Dordrecht, Holland

Sold and distributed in the U.S.A., Canada, and Mexico
by D. Reidel Publishing Company, Inc.
Lincoln Building, 160 Old Derby Street, Hingham,
Mass. 02043, U.S.A.

Printed in The Netherlands

TABLE OF CONTENTS

ACKNOWLEDGMENTS

This book is based on investigations which have spanned several years. A major part of this research was supported by National Science Foundation Grant GS-39664. This support made possible a leave from teaching and is greatly appreciated. I also appreciate useful supplemental funds provided by the University of Texas Research Institute.

Parts of this book are based on material from some of my previous publications. I wish to thank the editors and publishers for permission to reprint, in considerably revised form, parts of: 'Uniform Microreductions', *Synthese* **25** (1972); 'Attribute-Identities in Microreductions', *The Journal of Philosophy* **69** (1972); 'Unified Theories and Unified Science', in A. C. Michalos and R. S. Cohen (eds.), *PSA 1974*, D. Reidel, Dordrecht (1976); 'Laws, Identities, and Reduction', in M. Przełęcki, K. Szaniawski, R. Wójcicki (eds.), *Formal Methods in the Methodology of Empirical Sciences: Proceedings of Conference for Formal Methods, Warsaw, 1974*, Ossolineum and D. Reidel, Wrocław and Dordrecht, forthcoming.

Many persons have commented, both privately and in published articles, on my previous publications. These comments have been helpful in the writing of this book, and I thank collectively all who have contributed their comments. I also wish to thank Professor Jaakko Hintikka for his interest in this project, and to acknowledge his referee's useful comments on the first draft of the manuscript.

My wife, Sandy, has typed most of the final draft, and has assisted with the proofreading and the preparation of the figures. She has provided useful editorial comments, and I am extremely grateful for her constant encouragement.

The research reported in this book is, of course, my own responsibility. The analyses and methodological principles defended herein are not necessarily endorsed by any of the individuals or institutions mentioned above.

CHAPTER 1

INTRODUCTION

The first section of this chapter describes the major goals of this investigation and the general strategy of my presentation. The remaining three sections review some requisite background material and introduce some terminology and notation used in the book. Section B contains a brief review of some of the ideas and notation of elementary logic and set theory. Section C contains an introductory discussion of kinds and attributes. Section D presents some basic ideas about laws and law-sentences.

A. General plan of the book

Basic scientific research is directed towards the goals of increasing our knowledge of the world and our understanding of the world. Knowledge increases through the discovery and confirmation of facts and laws. Understanding results from the explanation of known facts and laws, and through the formulation of general, systematic theories. Other things being equal, we tend to feel that our understanding of a class of phenomena increases as we develop increasingly general and intuitively unified theories of that class of phenomena. It is therefore natural to consider the possibility of one very general, unified theory which, at least in principle, governs all known phenomena. The dream of obtaining such a theory, and the understanding that it would provide, has motivated an enormous amount of research by both scientists and philosophers.

Of course, such a unified theory is a very remote ideal at the present time, and it was an even more remote goal in the past. Nevertheless, there has long been interest in the possibility of a unified science which would provide a general understanding of the various classes of natural and social phenomena. Some important scientific advances have involved partial unifications of previously independent areas of scientific research. These advances have usually been accomplished by the reduction of one scientific theory to another. Philosophers have examined the prospects of a unified science in a number of ways. The Logical Positivists were concerned to defend a certain kind of very general program for the unification of science as presented in Neurath *et al.* (1938). In more recent years research in this field has become more specialized, involving studies of reduction, of functional and teleological theories, and of explanation in the various branches of science. One general study of the unity of science in recent years is that of Oppenheim and Putnam (1958). They discuss the possibility of the unification of science by means of successive microreductions of various theories to a basic theory about fundamental particles. This kind of reductionistic program faces many difficulties that have been widely discussed by a number of investigators, including the present writer in Causey (1968a). Nevertheless, it seems to me that a properly characterized reductionistic program offers the best overall plan for the unification of science. This book is devoted to a very detailed formulation and defense of a certain kind of reductionistic program for the unification of science. During the past few years there have been important scientific advances towards increased unification of the sciences. In addition, there have been significant developments in our understanding of scientific methodology. For these reasons I believe that the present time is especially appropriate for the formulation and defense of a detailed program for the unification of science.

It is convenient to make a rough division of scientific theories in either the natural or social sciences into three general types: (i) *dynamic theories*, which state and explain (as far as possible) the general laws governing the attributes and behavior of the various kinds of things in a specified domain; (ii) *developmental theories*, which describe and explain the general types of changes which take place over time in certain kinds of things under certain conditions; and (iii) *evolutionary theories*, which describe and explain actual changes which take place over time in some particular thing or class of things.

This is obviously a very rough classification of theories; yet, even from these very brief descriptions, it is clear that the structure of science is quite complex. Consequently, the unification of science will be a complex endeavor. The major part of this endeavor is the unification of the various dynamic theories; this book is primarily devoted to the development and defense of a program for the unification of these dynamic theories. In addition, a total unification program must take into account the developmental and evolutionary theories. I will therefore also include some limited discussion of these kinds of theories.

My principal discussions and arguments begin in Chapter 2, the first section of which contains a discussion of the logic of explanation. However, the primary aim of this chapter is to argue that a purely set theoretical language is inadequate for many important scientific purposes. In particular, it is argued that a purely set theoretical language does not provide an unequivocal linguistic representation of laws and explanations. For this and other reasons, it is proposed that the predicates in laws should normally be interpreted as denoting kinds or as denoting attributes. I do not attempt, in Chapter 2, to provide conclusive reasons for this mode of interpretation. Additional reasons for favoring this particular mode of interpretation are provided in later chapters. However, Chapter 2 does discuss certain consequences and applications of this interpretation of the predicates in laws. I state a criterion which determines whether or not two sentences stating laws state the same law. A somewhat similar criterion is also presented for the derivations which represent causal explanations. This chapter also introduces the important concept of a noncausal sentence, and it contains discussions of identities of kinds and of attributes. These types of identities play a fundamental role in the program for the unification of science. Chapter 2 therefore introduces much of the basic conceptual apparatus which is subsequently used in the formulation and defense of the unification program. This conceptual apparatus is novel and, I believe, intrinsically interesting. It provides new insights into the ontology of scientific theories. Among other things, the unification program involves the unification of the ontologies of various scientific theories. Chapter 2 concludes with a brief description of some very general features of dynamic theories. The specific conditions for explanations and theories which are presented in this chapter are quite skeletal. They are augmented with additional conditions in later chapters.

Chapters 3, 4, and 5 are concerned with the conditions for the reduction of one theory to another. Roughly speaking, a reduction of a theory T_2 to a theory T_1 involves explanations of the laws of T_2 in terms of the laws of T_1. However, many complex and subtle conditions must be satisfied in order to have an adequate reduction. In particular, I focus attention on the detailed conditions for microreductions. In these reductions the kinds of things in the domain of T_2 are generally considered to be structured wholes composed of parts which are elements of the domain of T_1.

Chapter 3 begins with a careful examination of the basic logical and semantical features of theories with structured wholes. Such theories have in their domains some elements which are considered simple and indecomposable from the point of view of the theory in question, and other elements which are structured wholes whose parts are various kinds of indecomposable elements. Theories of this kind play an essential role in what I call uniform microreductions. In Chapter 4 I present the conditions for uniform microreductions assuming, as far as is possible, that the theories involved utilize set theoretical languages. These conditions are useful and interesting, but they are too weak to provide an entirely adequate characterization of uniform microreductions. In Chapter 5 I then argue that set theoretical languages are not sufficient for the formulation of the complete set of conditions for adequate uniform microreductions. In particular, I argue that the connecting sentences relating the reduced theory to the reducing theory should all be identities. This is a strong condition which has interesting and useful consequences, which are also discussed. Chapters 3, 4, and 5 together constitute what I believe to be the most detailed discussion of microreductions ever presented. In addition, I believe that I present a convincing argument that the set of connecting sentences in uniform microreductions should consist of identities of kinds plus identities of attributes. Needless to say, this conclusion provides further support for my general proposal that the predicates in laws be interpreted as denoting kinds or attributes.

The idea that we should try to unify theories by means of microreductions is a very natural one. Yet, there has been little systematic study of the characteristics of unified theories, and little foundational support for the use of reductions as a unifying procedure. Moreover, even if all dynamic theories were reduced to some fundamental theory, we would still desire in addition that this fundamental theory be a *unified theory*. Therefore, in Chapter 6 I state and defend some conditions which are necessary in order for a theory to be unified. I then argue that, when one theory can be microreduced to another, then the only satisfactory way to combine these two theories into one unified theory is by a microreduction. Chapter 6 thus provides the foundational justification for the use of microreductions in a program of unification, and it also provides at least some of the necessary conditions for a unified theory.

Chapters 2 through 6 present the general reduction conditions and the general rationale for an overall program for the unification of dynamic scientific theories by means of successive microreductions. Basically, the *unification program* consists of the unification of all branches of science by means of successive microreductions of the dynamic theories of these branches to one unified theory. This is a very general research program which will encounter various types of complications and obstacles. In recent years there have been many arguments by philosophers and scientists for or against the possibility of the reduction of certain types of theories to certain other types of theories. These arguments frequently rely on rather specialized metaphysical or empirical assumptions, for instance, assumptions concerning relationships between mental and physical phenomena. In the present investigation I almost entirely avoid discussing these specific controversial issues. I am concerned, however, by the fact that many of the discussions of these issues have suffered from the use of inadequate reduction conditions or the use of quite over-simplified programs for the unification of science. The present investigation should help to illuminate some of these issues by providing a general characterization and defense of a nontrivial and realistic unification program.

Although I wish to avoid specialized controversies concerning the reducibility of particular theories, nevertheless, it is important to consider the principal, general types of complications and obstacles facing the unification program. Chapter 7 is devoted to this task. I first discuss some complications resulting from the enormous variety of types of structures found in the world, and then I consider some special problems encountered in connection with theories about hierarchical structures. The remainder of this chapter is primarily devoted to a consideration of some complications involved in the reduction of the social to the physical sciences. Obviously, Chapter 7 is not intended to provide an exhaustive discussion of all possible types of obstacles facing the unification program. However, it does contain considerations of some of the most troublesome obstacles, and I believe that I show that they are not necessarily insuperable.

One might justifiably question the significance of having unified theories or a unified science. It would be nice to be able to provide detailed motivations for these aims in this first chapter. Unfortunately, this is not feasible because it is necessary to present first the detailed characterization of the unification program. Consequently, I delay until Chapter 8 the systematic presentation of the motivations for the unification program. Chapter 8 contains a discussion of some very important aspects of scientific progress. This progress involves increases in both knowledge and understanding of the world. The unification program is especially connected with our understanding of the world. In Chapter 8 I describe a number of ways by which unified theories and the reduction of theories can enhance our scientific understanding. Thus, it is argued that the unification program presented in this book provides certain conditions which must be satisfied in order to obtain a certain type of understanding of the world. The program therefore provides certain conditions which must be satisfied in order for scientific research to progress in certain ways, and this in turn provides a plan for scientific research.

Of course, scientific research and scientific progress do not only involve dynamic theories. For this reason, Chapter 8 also contains some discussion of the role of developmental and evolutionary theories in scientific progress. I then show how the unification program (for dynamic theories) is related to scientific progress in the study of developmental and evolutionary processes. It turns out that explanations of developmental processes can be absorbed satisfactorily into appropriate dynamic theories. On the other hand, the relationships between dynamic theories and evolutionary theories are more complex and quite interesting. Evolutionary theories do not directly fit into the unification program; however, if the program is successful, it will impose certain constraints on acceptable evolutionary theories. These constraints can be helpful in the construction of evolutionary theories.

I wish to emphasize that the unification of science is an ideal which will probably never be totally achieved. Indeed, the achievement of any reduction, or the construction of any unified theory, depends in part on the nature of the real world. Thus, there can be empirical barriers preventing the total unification of science, just as there can be empirical barriers preventing the accomplishment of certain other aspects of scientific progress. In the last section of Chapter 8 I make a few remarks concerning some additional problems and prospects connected with the unification program. Some of these remarks are concerned with possible empirical limitations on scientific progress and some are concerned with more general philosophical issues. Nevertheless, I remain optimistic that scientific progress will continue, and that the unification program will play a very significant role in this progress. I recognize that the unification program faces obstacles and that alternative conceptions of scientific progress can be proposed. Unified science is an ideal towards which we can strive; the nearer we approach this ideal, the deeper will be our understanding of the world.

This entire book contains an ordered sequence of arguments. Each chapter depends upon and builds upon results of all earlier chapters. Therefore, the chapters must be carefully read in the order in which they occur in the book. In addition, this book is not elementary. It presumes some background in logic and scientific methodology. Nevertheless, it should be understandable by scientists and philosophers who are familiar with the basic characteristics of laws and theories. In the remaining sections of this chapter I will briefly review some of the general background knowledge which is presupposed.

B. Sets and notation

Some elementary logic and set theory will be used in this study. I assume that the reader is familiar with informal set theory and with first-order logic, including some model theory. On the whole, my discussions will be informal, although a general familiarity with axiomatic systems will be assumed. I also assume familiarity with some concepts of algebra, elementary mathematical analysis, and elementary probability theory. In this section I will briefly describe some of the notation which will be used.

Suppose that we have a nonempty set of objects, the *domain*, which is denoted by '*Dom*'. Let A, B be subsets of *Dom*, and let a be an element of *Dom*. Then, as is usual, '$a \in A$' means that a is a member (element) of A, '$A \subseteq B$' means that A is a subset of B, and '$A \cup B$', '$A \cap B$' denote, respectively, the union and intersection of A and B. If $a_1, \ldots, a_n$ are members of *Dom*, then $\{a_1, \ldots, a_n\}$ is the set containing exactly $a_1, \ldots, a_n$, and $\langle a_1, \ldots, a_n \rangle$ is the ordered n-tuple with first element a_1, second element a_2, and so on. A^n is the set of all ordered n-tuples of elements in A.

A set of ordered n-tuples of elements of *Dom* is normally called a 'relation on *Dom*'. For reasons to be stated in Section C, I will not use this terminology. Instead, a set of ordered n-tuples of elements of *Dom* will be called an 's-relation on *Dom*', or simply an 's-relation'.

It will be convenient to use some notation of symbolic logic, although my later interpretation of this notation will be partly nonstandard. If 'x' is an individual variable, then '(x)', '$(\exists x)$' denote, respectively, universal and existential quantification with respect to 'x'. Negation is denoted by '$-$', disjunction by '$\vee$', and conjunction by '$\&$'. I use '$\rightarrow$' for the material conditional, and '$\leftrightarrow$' for the material biconditional. I will use various symbols for predicates depending on the subject matter discussed. My interpretation of predicates will be discussed in the next section.

As is well known, first-order logic is weak, and it is insufficient for the representation of many mathematical and scientific theories. However, by using the standard interpretation of '∈', set theory can be axiomatized in first-order logic, and most interesting mathematical theories can be developed within set theory. Nevertheless, in the course of this work I will present various reasons for the inadequacy of set theory for many scientific purposes. Therefore, I will not assume that scientific theories can be adequately formalized in set theory or any other known formal language. Consequently, I will use symbolic formulations informally as is done in ordinary mathematical and scientific writing. Likewise, when I use the terms 'derivation' or 'deduction', I will normally be referring to arguments presented informally according to the usual standards of sound mathematical and scientific reasoning.

C. KINDS AND ATTRIBUTES

Our ordinary language contains many proper names or individual constants which denote particular persons, places or things. Examples of such terms are 'Romeo', 'Juliet', 'Mt. Everest', 'London'. We also use general terms such as 'raven', 'black', 'snow', 'gold'. Some of these terms are ordinarily thought of as names for *kinds* of things, and others are interpreted as names of *properties*. In addition, we use terms designating *relations* and *quantities*. In the course of this work I will have much to say about kinds, properties, relations, and quantities. Any property, relation, or quantity will be considered an *attribute*. In the present chapter I will discuss kinds and attributes informally on an intuitive level; various reasons for the use of these concepts will emerge in later chapters.

A predicate which denotes a kind of thing is a *thing-predicate*. Examples of thing-predicates are 'raven', 'snow', 'gold', 'hydrogen atom', 'light wave', 'water'. Some thing-predicates are called 'sortal predicates' because they apply to things which can be individuated and counted discretely in a definite way; see, for instance, Strawson (1963, p. 169). The *extension* of such a predicate is simply the set of all objects to which the predicate truly applies. For instance, the extension of 'raven' is just the set of all ravens, and the extension of 'hydrogen atom' is the set of all hydrogen atoms. I will assume that these sets exist at any given point in time.

Some thing-predicates are called 'mass terms' because objects which satisfy them can be divided or combined in various ways to form new objects which satisfy them. For example, two samples of water can be combined to form one sample, and a gold bar can be cut into two samples of gold. Many subtle problems arise in the interpretation of mass terms as used in ordinary language. The interested reader may consult Quine (1960, p. 91) and several specialized articles in Pelletier (1975). For the purposes of the present investigation a simple interpretation of mass terms will suffice. Unless otherwise indicated, I will follow the convention that the extension of a mass thing-predicate is, at any given time, the set of all objects which, at that time, satisfy that thing-predicate.

Scientific theories use, in addition to thing-predicates, certain *attribute-predicates*, which denote attributes. Some attribute-predicates denote properties such as white, yellow, spherical. The attribute-predicate 'white', which denotes the property of being white, also has an extension, namely, the set of all white things. Similarly, the extension of 'spherical' is the set of all things which are spherical. I assume that these extensions exist at any given time.

Some attribute-predicates denote relations such as brother of, and between. Let us denote the set of all people by Dom, and introduce the following symbols: 'Fx' means x is a female, 'Bxy' means x is a brother of y, '$Txyz$' means y is between x and z, and 'a','b','c' denote three particular people. Then 'Bab' is a sentence which states that a is a brother of b, 'Fb' states that b is a female, and '$Tabc$' states that b is between a and c. In general, if ϕ is any predicate which denotes a kind, property, or relation, 'Ext_ϕ' denotes the extension of ϕ. Thus, Ext_F is the subset of Dom which consists of all females, Ext_B is the subset of Dom^2 which consists of all ordered pairs $\langle x, y \rangle$ such that x is a brother of y, and Ext_T is the subset of Dom^3 which consists of all ordered triples $\langle x, y, z \rangle$ such that y is between x and z. Ext_B and Ext_T are s-relations on Dom. Of course, Fb implies that $b \in Ext_F$ and $Tabc$ implies that $\langle a, b, c \rangle \in Ext_T$. In so-called extensional logics, predicates such as 'F' and 'T' are simply interpreted as denoting their extensions, Ext_F and Ext_T. Extensional interpretations are adequate for some purposes, but not adequate for all scientific purposes, as I will show in later chapters. Therefore, except when otherwise indicated, I will continue in an intuitive manner to use interpretations in terms of kinds, properties, and relations. It must be remembered that in my terminology a relation is an attribute which has an s-relation as its extension. Properties can be thought of as 1-place relations, and sets as 1-place s-relations.

If we are given certain predicates, it is possible to define other predicates in terms of the given ones. For instance, if we have 'white' and 'spherical', we can define the attribute-predicate 'white and spherical'. If we are given the 'F' and 'B' above, we can define '$(\exists y)(Fy\ \&\ Bxy)$' which means that there is a female y such that x is a brother of y. I assume that the reader is familiar with the usual adequacy requirements for definitions in extensional languages; one may consult Suppes (1957, Chapter 8) for details. There are also further, theory-dependent, restrictions on the kinds of combinations of predicates which are acceptable as defined thing- and attribute-predicates. These restrictions will be discussed in later chapters.

We must now consider another class of attributes, namely, quantities. First, however, consider the property of being spherical. An object is either spherical or it is not spherical. We could assign the numbers 0 or 1 to objects according to whether they are not or whether they are spherical. At least in sharply defined applications, ordinary properties are applied in a two-valued manner. Intuitively, we usually think of a quantity, say mass, as a kind of property which is possessed by objects to various degrees, where more than two degrees are possible. Because of this a quantity must be measurable by various procedures which enable us to assign numerical values to objects according to the degree to which they possess this quantity. The branch of methodology called 'the theory of measurement' is concerned with the conditions and procedures for the measurement of various types of quantities.

If the measurement of a quantity Q makes use of previously given quantities, then Q is usually said to be a 'derived quantity'. The theory of derived measurement is complex and will not be discussed here in any detail; recent theories of derived measurement can be found in Causey (1967), Causey (1969a), Luce (1971), Krantz *et al.* (1971, Chapter 10). Some quantities are measured without the use of any previously given quantities and are therefore called 'fundamental'. Fundamental measurement is accomplished with the help of certain empirically given relations and operations on a set of objects. Let us briefly review an example of a theory of fundamental measurement which is presented in Suppes and Zinnes (1963, pp. 41–45).

Consider an s-relational system $\langle Dom, R, \circ \rangle$, where R is a binary s-relation on *Dom*, and $\circ$ is a function mapping Dom^2 into *Dom*. Assume that for every x, y, z in *Dom*, $\langle Dom, R, \circ \rangle$ satisfies

$$Rxy\ \&\ Ryz \rightarrow Rxz \qquad\qquad (1.1)$$

$$R((x \circ y) \circ z)(x \circ (y \circ z)) \tag{1.2}$$

$$Rxy \to R(x \circ z)(z \circ y) \tag{1.3}$$

$$-Rxy \to (\exists z)(Rx(y \circ z) \ \& \ R(y \circ z)x) \tag{1.4}$$

$$-R(x \circ y)x \tag{1.5}$$

$$Rxy \to (\exists n)Ry(nx) \tag{1.6}$$

where nx is defined recursively as follows: $1x = x$ and $nx = (n-1)x \circ x$. $\langle Dom, R, \circ \rangle$ is considered to be a possible empirical *s-relational system*.

Using R, one can define an indifference s-relation I on *Dom* such that I is an equivalence s-relation, i.e., I is reflexive, symmetric, and transitive. Hence, I partitions *Dom* into disjoint equivalence classes, and we can consider the new s-relational system $\langle Dom/I, R^*, \circ^* \rangle$ modulo I, where R^* and $\circ^*$ are defined on *Dom/I* and correspond to R and $\circ$ in the natural way.

Now let $\langle N, \leqslant, + \rangle$ be an s-relational system where N is a non-empty set of positive real numbers closed under addition and closed under subtraction of smaller numbers from larger numbers, where $\leqslant$ is equal to or less than, and where $+$ is addition. Such a system is a *numerical extensive system*.

It is now possible to prove two important theorems. One first proves a *representation theorem* which states that $\langle Dom/I, R^*, \circ^* \rangle$ is isomorphic to some numerical extensive system. It follows that $\langle Dom, R, \circ \rangle$ is homomorphic to this numerical extensive system, and a function which provides such a homomorphic mapping is a *representation function*. However, there is not a unique representation function, so it is desirable to determine the class of all such functions. One therefore proves a *uniqueness theorem* which states, in this case, that any two representation functions f_1, f_2 are related by a *similarity transformation*, i.e., there is a positive real number r such that, for any x in *Dom*, $f_2(x) = rf_1(x)$.

To illustrate an application of this abstract analysis, suppose that we have a balance scale capable of holding the objects in *Dom*. For any x, y in *Dom*, Rxy if and only if the balance indicates that x is less heavy than y, or x is equally heavy as y. For any x, y in *Dom*, $x \circ y$ is the result of combining x and y on the same pan of the balance. This interpretation, plus some idealizing assumptions covering limiting empirical cases, provides an empirical realization of (1.1)–(1.6). The representation theorem then shows that the *masses* (or, if preferred, *weights*) of the objects in *Dom* can be measured. The uniqueness theorem describes the kind of relationship which holds between any two given representation functions for mass measurement. A study of the proof of the representation theorem shows that a particular representation function for mass measurement is determined by fixing the measurement of some chosen object as having the value 1. For example, a certain object may be used as the standard kilogram to determine masses in the metric system; another object may be used as the standard pound to determine masses in pounds.

Measurement theorists have studied a large variety of different kinds of s-relational systems. They have proved many representation and uniqueness theorems. As is well known, measurements of mass are said to be made on a *ratio scale* because any two representation functions are related by a similarity transformation. There are other kinds of measurements which utilize representation functions which are related by other kinds of mathematical transformations. In addition, the scale type of a derived measurement is usually determined as a function of the scale transformations of the measurements in terms of which it is derived. I am not concerned with such technicalities here, for the discussion just given should help make clear certain distinctions which will be introduced later in my treatment of quantities. For the present, it should be realized that an adequate understanding of a statement such as "The height of Mt. Everest is 8847.7 m" presupposes a great deal. In particular, we must understand that the reference to meters involves understanding that the number 8847.7 is the numerical value assigned to Mt. Everest by a particular representation function for the measurement of lengths (and distances). As will be seen below, it is customary to state laws involving quantities in a more general way which does not refer to any particular representation function. The way this is accomplished can be illustrated by a simple statement which is not a law.

Suppose that we have two objects a, b such that b weighs 2 kg and a weighs 1 kg. Let f be the function which assigns to objects their masses in kilograms. Then

$$f(b) = 2f(a). \tag{1.7}$$

But any two representation functions for mass are related by a similarity transformation. Therefore, if m is any representation function for measurement of mass, we have

$$m(b) = 2m(a). \tag{1.8}$$

Corresponding to (1.8), we would ordinarily say, "The mass of b is twice the mass of a." We are justified in making such a statement because of the uniqueness theorem concerning measurement of mass.

Of course, balance scales are not the only devices for measuring masses, and so it must *not* be thought that the representation functions for mass measurement are operationally defined in terms of the particular empirical realization described above. Given two systems of measurement, we can ask whether they provide measurements of the same or different quantities. Conditions for the identity of quantities will be discussed in Chapter 2, Section B.

D. LAWS AND LAW-SENTENCES

Natural laws are certain types of true and general statements. I will say that a law is expressed or stated by a *law-sentence*. This allows the possibility, which will be discussed in Chapter 2, that different law-sentences may state the same law. I assume that the reader is familiar with modern philosophical analyses of laws, and so in this section I will only briefly describe certain aspects of laws which are especially relevant to this study.

1. Some law-sentences are qualitative, and so use no names for quantities. The simplest form for a qualitative law-sentence is a universally quantified conditional of the form

$$(x)(\alpha x \to \beta x) \tag{1.9}$$

whore α and β are 1-place predicates. Examples of law-sentences of form (1.9) are: "All ravens are black," which means "For any x, if x is a raven, then x is black," and "Gold is malleable," which means "For any x, if x is a sample of gold, then x is malleable."

Qualitative law-sentences may have forms more complex than that of (1.9), for example, there is the law-sentence, "For any x and any y, if x is a sample of diamond and y is a sample of talc, then x is harder than y." Universal law-sentences must begin with a universal quantifier, but they may also contain existential quantifiers and other logical connectives.

2. Many laws are quantitative. Consider the well-known ideal gas law, usually expressed by

$$pV = nRT, \tag{1.10}$$

where p is pressure, V is volume, n is the number of moles of gas, R is a universal constant whose numerical value depends only on the units of measurement used, and T is the absolute temperature of the gas. (1.10) is only a partial statement of the law; a more precise statement of this law can be formulated as follows: "For any x, n, p, V, T, if x is a sample of a gas, n is a representation function for moles, p is a representation function for pressure, V is a representation function for volume, and T is a representation function for absolute temperature, then there is a constant R, whose value depends only on the choice of representation functions, such that $p(x)V(x) = n(x)RT(x)$." In addition, since some of the quantities involved in this statement are derived quantities, it is understood that the representation functions use a so-called 'coherent system of units'. This qualification can be taken for granted in the present context.

The ideal gas law is presented here in order to illustrate the complexity of quantitative laws. Of course, it is a relatively simple quantitative law. Many others are much more complex and utilize such mathematical devices as differential equations, matrix algebra, infinite series, and so on. In addition, most quantitative laws are known to be only approximately true in certain circumstances. Sometimes they can be replaced by more accurate laws. Fortunately, most of these complications will not concern us, at least in any detail. But it is important to know how to convert abbreviated forms of quantitative law-sentences (such as (1.10)) into their more precise statements.

3. Some laws have the form of a universally quantified biconditional, namely,

$$(x)(\alpha x \leftrightarrow \beta x), \tag{1.11}$$

where α and β are predicates. When (1.11) holds, then also $Ext_\alpha = Ext_\beta$. For this reason, a law of the form (1.11) may be called a '*nomological co-extensionality*'. Laws of this form are also frequently called '*nomological correlations*', or simply '*correlations*'. A simple example of a nomological co-extensionality is the following: Let '*sink x*' mean x has the disposition to sink in H_2O in a gravitational field, and let '*den(x)*' denote the density of x. Then

$$(x)(sink\ x \leftrightarrow den(x) > den(H_2O)). \tag{1.12}$$

There are many nomological co-extensionalities in addition to (1.12). Laws of this form will play an important role in later discussions.

4. Many important laws are not strict universal generalizations, but are instead probabilistic statements. One simple form of a probabilistic law-sentence is

$$pr(\alpha, \beta) = r, \tag{1.13}$$

where $pr(\alpha, \beta)$ is the probability that an object x will have the property α, given that x has the property β, or is of the kind β. In (1.13), r is a real number such that $0 \leqslant r \leqslant 1$. Although there are different interpretations of the mathematical theory of probability, the most usual interpretation placed on laws of the form (1.13) is relative frequency. Under this interpretation, r is the relative frequency of α's among β's. This is the interpretation I will adopt. For example, the frequency of albinism in humans is 1/10,000; see Bearn (1971, p. 1698). This fact can be stated as a law of the form (1.13).

5. It is customary to distinguish genuine laws from so-called accidental generalizations. Borrowing an example from Hempel (1966, p. 55), suppose that it just happens to be true throughout the entire life-span of the universe that all solid, spherical samples of gold have a mass less than 10^6 kg. This would be a curious fact which is stated in the form of a universal generalization. Yet, it would not normally be considered a law. Instead, it would be considered an accidental circumstance. I think that the principal reason we would consider it an accident is the fact that we accept other laws and theories which imply that it could be falsified. In other words, our currently accepted and well-confirmed theories of matter imply that it is physically possible to construct a solid sphere of gold with mass greater than 10^6 kg.

By way of contrast, consider the claim that all solid, spherical samples of ^{235}U have mass less than 10^2 kg. This claim has the same general semantical structure as the previous assertion about gold. Yet, the claim about ^{235}U is not only true, but it is a law. One can now find in standard references such as Lapedes (1971, p. 602) that the critical mass of ^{235}U is about 16 kg. By the laws of nuclear physics it follows that a solid sphere of ^{235}U weighing more than 10^2 kg will undergo a spontaneous fission chain reaction. Under suitable conditions it will be an atomic bomb.

These examples illustrate that the syntactical and semantical structures of sentences cannot be relied upon to distinguish genuine laws from merely accidental universal generalizations. However, it is widely accepted that a universal generalization is a law-sentence if it is casually explainable within an accepted theory from a set of premises consisting only of other law-sentences of this theory. This ultimately requires that some set of sentences must be accepted as fundamental law-sentences of this theory, and there may be no sure test to show that they are not merely accidentally true generalizations.

The accidental/nomological distinction produces analogous, and perhaps even more severe, problems in the interpretation of statistical generalizations expressed in the form of (1.13). Fortunately, in this work I will be mainly concerned with various kinds of relations between sentences which are accepted as laws, and so the accidental/nomological distinction will not play any significant role in the analyses or arguments that follow. For further information concerning this distinction, the reader is advised to consult other sources, for instance, Nagel (1961, Chapter 4). In addition, a very useful discussion of probabilistic laws can be found in Mackie (1974, Chapter 9). *Henceforth, all of the principles and arguments presented in this book will be concerned with nonaccidentally true sentences unless specifically stated otherwise.*

One final remark must be made here concerning the word 'correlation'. Suppose that we have confirmed an empirically true co-extensionality with the syntactical form of (1.11). Without any further information, the proper interpretation of this co-extensionality will remain open. However, depending on additional considerations, we might decide that this co-extensionality is an accidental generalization or that it is a nomological co-extensionality. As I will show later, we might even decide that it represents an identity of kinds or of attributes. Unfortunately, the word 'correlation' has been used by various writers in different ways. In order to avoid confusion, I will use the unmodified word 'correlation' to mean 'nomological co-extensionality'. If any other sense of 'correlation' is intended, I will explicitly state the appropriate qualification.

CHAPTER 2

EXPLANATIONS, IDENTITIES, AND THEORIES

In the first section of this chapter I will discuss certain aspects of the logical structure of explanations. The second section will continue the discussion of kinds and attributes, which was begun in Chapter 1. I will show how to distinguish synthetic thing-identities and attribute-identities from nomological co-extensionalities. This section will also examine conditions under which different law-sentences express the same law. Finally, in Section C, I will describe some general features of the structure of scientific theories, and I will state a sufficient condition for two explanatory derivations to represent the same explanation.

A. Scientific explanations

Basic scientific research has two general goals: to increase our knowledge of the world and to increase our understanding of the world. We increase our knowledge by discovering new particular facts and new general laws. We increase our understanding mainly by constructing explanations of previously discovered knowledge, and by organizing the knowledge and explanations into systematic and general theories. Knowledge and understanding are related in many ways, but they are not the same. For example, it was known *that* the planets move in elliptical orbits before it was understood *why* they move in elliptical orbits. In general, an explanation is an answer to a *why-question*, e.g., "Why do the planets move in elliptical orbits?", "Why does water expand when cooled to $4° C$?", "Why does infection with polio virus sometimes cause paralysis?". If someone asks, "Why is such and such the case?", we try to answer by giving a reason why such and such is the case. In a teleological explanation, the reason is a *purpose* for such and such. In other cases the reason for such and such is given in the form of an argument. If one accepts the premises of the argument and agrees that the argument is sound, then he is supposed to understand why such and such is the case in terms of the information provided by the premises of the argument. In this section I will briefly consider such explanatory arguments as used in order to provide scientific explanations.

There is a large body of literature, and there has been much controversy, about the logical structure of scientific explanations. I do not intend to review this literature in any detail. Fortunately, on the whole the present investigations require assuming only some of the more general and less controversial features of explanations. In this section I will describe a few examples of explanations in order to illustrate what I assume in general about explanations, and why.

1. The most familiar form of explanatory argument is that used in so-called *deductive-nomological* (D-N) explanations. Following Hempel (1965, pp. 336–337), a *D-N derivation* is usually said to have the general form

$$\frac{\begin{array}{l} L_1, L_2, \ldots, L_m \\ C_1, C_2, \ldots, C_n \end{array}}{E}$$

where the set of premises, $\{L_1, \ldots, L_m, C_1, \ldots, C_n\}$, is the *explanans*, and the conclusion E is the *explanandum*. In addition, certain assumptions must be made about the sentences involved. My assumptions are somewhat more general than those usually found in the literature, and I will generalize the entire form of the argument even further as the discussion proceeds. For the present, I require that: (i) the L_i are law-sentences; (ii) the C_j (the *boundary conditions*) are sentences describing particular or generic events, or particular or generic states-of-affairs; (iii) E describes a particular or generic event or state-of-affairs, or is a law-sentence; (iv) the sentences in the explanans are true; (v) the derivation is a logically sound derivation of E from the explanans; (vi) the explanans contains at least one law-sentence required for the derivation (and there may be no C_j).

A few remarks are now in order. First of all, it should be clear that one will rarely find scientific papers or texts with explanatory arguments stated precisely in the form above. However, I do believe that good, causal explanations can be formulated in this form (or in the generalized forms to be presented below) when proper attention is given to all assumptions made. Very generally, the reason for requiring a law-sentence in the explanans is because causal connections are typically expressed by laws, and because D-N derivations are generally supposed to represent causal explanations. This will become clearer when we look at some examples.

There has been considerable controversy over condition (iv). Some would prefer to replace the word 'true' by other phrases such as 'well-confirmed'. There may be some justification for such replacements when one is considering the conditions for the practical acceptability of explanatory arguments. However, it will become clear later that 'true' is both sufficient and appropriate for the investigations to be presented in this book.

Finally, as I have indicated in Chapter 1, by a 'logically sound derivation' I mean an argument which is sound according to the usual standards of mathematical and scientific reasoning. I do not assume that any particular type of formal system of logic is used in D-N derivations. Let us now consider some examples.

Suppose that we dissolve some silver nitrate, $AgNO_3$, in H_2O. Two platinum electrodes are placed in the solution and connected to a direct current source. Soon silver metal appears on one of the electrodes. Why does silver metal form? Formulated rather crudely, the explanation is as follows: (L_1) When $AgNO_3$ is dissolved in H_2O it forms Ag^+ and NO_3^- ions. (L_2) When there is a negative electrode in a solution containing Ag^+ ions, these ions are attracted to this negative electrode. (L_3) At a negative electrode in a solution containing Ag^+ ions, the Ag^+ ions react with electrons to form Ag metal. (C_1) $AgNO_3$ is dissolved in H_2O. (C_2) Two platinum electrodes, one charged negatively and one charged positively by a direct current source, are placed in the silver nitrate solution. Therefore, (E) silver metal forms.

This example is over-simplified from both a logical and a scientific standpoint. However, it is sufficiently detailed to illustrate how a D-N derivation explaining a particular event involves an interplay between law-sentences and sentences describing particular events and conditions. In this example, the C_j describe particular events or conditions. However, these C_j and E could be interpreted as describing events and conditions of a general kind. Then from L_1 & L_2 & L_3, we could derive: "If conditions of type C_1 and C_2 obtain, then E occurs." This is a specialized *derivative law-sentence* which is explained by the law-sentences L_1, L_2, L_3. Thus, we also have an example of a D-N derivation which uses only law-sentences in its explanans. We will see later that explanatory derivations within theories usually do not contain C_j's.

2. A *logical truth* is a sentence that can be derived from the empty set of premises using sound rules of derivation. Let Γ be a set of premises that are not logical truths, and suppose that we form a new set of

premises $\Gamma \cup \Delta$, where Δ is a set of logical truths. Finally, suppose that we have a derivation of a sentence ϕ from $\Gamma \cup \Delta$. Then it is easy to change this derivation into a derivation of ϕ from Γ by starting with Γ, and, wherever needed, interpolating into our derivation from Γ derivations of the logical truths in Δ. Therefore, adding logical truths to the premises of a derivation is, in effect, not adding any new premises at all. This is an elementary principle of logic.

If we have a language which already contains certain terms, it is often convenient to define new terms using definitions utilizing the original terms. When the definitions are properly formulated, the defined terms are eliminable from any derivation. It is sometimes convenient to use such definitions as premises in derivations. When they are so used, they are treated as true sentences. However, they will be considered to be *analytically* true, and they are called '*analytic sentences*' because they are eliminable. In other words, they are true by the conventions expressed by their definitions, but these conventions are actually unnecessary and are used merely for convenience.

Actually, it is customary to use more general terminology according to which logical truths are considered to be analytic. I will follow this custom to this extent: Logical truths and definitions are *analytic*, and any sentence which is derivable from premises consisting only of logical truths and definitions is also *analytic*. From the remarks just made about derivations, we see that any analytic sentence can be added to the premises of a derivation without changing the premises of that derivation in any logically essential respect. This suggests that we generalize the conditions for a D-N derivation by allowing analytic sentences in the explanans. Thus, the explanans will have the form $\{A_1, \ldots, A_k, L_1, \ldots, L_m, C_1, \ldots, C_n\}$, where the A_j are analytic. The following example illustrates the use of this generalization, as well as some other features of D-N derivations.

Suppose that a body is dropped from a height above, but relatively near, the surface of the earth. It falls without friction to a lower height, at which point it has a certain velocity. We are given the following notation: γ is the universal gravitation constant, g is a constant which is a measure of acceleration, h is the height of the body above the surface of the earth at time zero, x is the height of the body at any time zero or later, r is the distance of the body from the center of the earth at any time, v is the velocity of the body at any time, $v_0 = 0$ is the velocity of the body at time zero, R is the radius of the earth, m, M are the masses of the body and of the earth, respectively, V is the potential energy of the body at any time, T is the kinetic energy of the body at any time, and V_0 is the initial potential energy of the body. We now wish to explain why

$$v = -[2g(h - x)]^{1/2}. \tag{2.1}$$

In (2.1) I use '$-$' for the minus sign; it should not be confused with logical negation. The minus sign occurs in (2.1) merely because it is assumed that measurements of distance are positive *away* from the center of the earth, and the body is falling *towards* the center of the earth. It should also be noticed that I am using the usual elliptical manner of stating quantitative laws, as was described in Chapter 1, Section C. This practice will be followed in describing the explanatory derivation.

In this derivation I assume that the term T for kinetic energy is defined, so we have the analytic premise

$$T = \tfrac{1}{2}mv^2. \tag{2.2}$$

We also assume the following two laws

$$V = -\gamma mMr^{-1}, \tag{2.3}$$

$$V + T \text{ is constant.} \tag{2.4}$$

(2.3) is the law which gives the value of the gravitational potential energy of a body with respect to the earth, and (2.4) is a case of the law of conservation of energy.

We now assume the following additional conditions

$$v_0 = 0, \tag{2.5}$$

$$r = R + x. \tag{2.6}$$

The initial distance of the body from the center of the earth is $R + h$, so the initial potential energy by (2.3) is $-\gamma mM(R + h)^{-1}$. By (2.5) and (2.2), the initial kinetic energy is zero. Hence, by (2.4)

$$V + T = -\gamma mM(h + R)^{-1}. \tag{2.7}$$

Combining (2.2), (2.3), (2.6), and (2.7), we obtain

$$\tfrac{1}{2}mv^2 - \gamma mM(R + x)^{-1} = -\gamma mM(h + R)^{-1}. \tag{2.8}$$

By applying some elementary algebra to (2.8), we find the following solution for v,

$$v = -\{2\gamma M(h - x)[(R + x)(h + R)]^{-1}\}^{1/2}. \tag{2.9}$$

Although (2.9) has been derived by a D-N derivation from the explanans, (2.9) appears considerably different from (2.1), our original explanandum. However, in fact the two sentences are closely related under the conditions that we have assumed. Since the body was dropped from an initial height relatively near the surface of the earth, and since the earth's radius is relatively large, $R + x$ and $R + h$ are both approximately equal in magnitude to R. If these approximations are imposed on (2.9), we obtain

$$v = -[(2\gamma M/R^2)(h - x)]^{1/2}, \tag{2.10}$$

where $\gamma M/R^2$ is a constant which is a measure of acceleration, and where it is understood that (2.10) is to be used only in cases where h is much less than R. In fact, it is easy to see that the right side of (2.9) approaches the right side of (2.10) as a limit as h/R approaches 0.

It will be noticed, however, that even after making the approximation or taking the limit, (2.10) is still not our original explanandum (2.1). However, (2.1) is derivable from (2.10) with the help of one further premise which is now added to our explanans,

$$g = \gamma M/R^2. \tag{2.11}$$

This premise is certainly not an analytic sentence; it is, in fact, a synthetic identity of quantities. Such identities will be discussed below in considerable detail. It will be immediately noticed, however, that (2.11) is a fourth kind of premise which may occur in the explanans of a D-N derivation. Unlike analytic sentences, (2.11) is a synthetic sentence which is subject to, and requires, empirical justification. Let us assume that it is justified. Then it allows the derivation of (2.1), and, in addition, it identifies g in terms of other constants occurring in the law-sentences of our explanans. When explaining a specialized law, such as (2.1), containing a specialized constant, such as 'g', scientists generally desire that the explanation produce an identification of this specialized constant in terms of constants occurring in the law-sentences or boundary conditions of the explanans. Moreover, no nontrivial logically sound derivation can have a substantive constant occurring in the conclusion which does not somewhere occur in the premises.

In this example we originally sought to derive a certain explanandum E, which was in this case (2.1). Instead after using what appeared to be the appropriate explanans, we derived (2.9). At this point we could have introduced the extra premise (2.11), and then obtained from (2.9) and (2.11)

$$v = - \left[2g(h - x)\left(1 + \frac{x}{R}\right)^{-1}\left(1 + \frac{h}{R}\right)^{-1}\right]^{1/2}. \tag{2.12}$$

Let (2.12) be E^*. Then we have explained the explanandum E^*, and E^* is approximately equivalent to E. This type of result can often be obtained, even if, considered precisely, E contradicts the explanans.

It might be claimed that we have only derived E^*, and so according to the conditions for D-N derivations, we have only explained E^*. However, since E is a good approximation to E^*, it makes good sense to say that we have given an *approximative D-N explanation* of E; see Hempel (1965, p. 344). Of course, it would be nearly impossible to try to characterize all kinds of allowable approximations. I only wish to point out that I will take for granted in my later discussions that approximations are often made in D-N derivations. This is allowable if it can be shown that the original explanandum E is approximately equivalent to the derived explanandum E^*. In the case of quantitative laws, one common way of showing this is by showing that E^* approaches E as a limit under appropriate conditions. Other ways of establishing approximate equivalence may be allowable in various contexts.

At this point brief mention should be made of deductive-statistical (D-S) derivations. It has been stipulated that the L_i in a D-N derivation are law-sentences. Usually these are restricted to universal law-sentences; however, I have not made this restriction. Therefore, we can consider a *D-S derivation* to be a special kind of D-N derivation satisfying the condition that the explanandum is a statistical (probabilistic) law-sentence. This will require, in order for the derivation to be possible, that the explanans contains at least one statistical law-sentence used in the derivation. D-S derivations may use the mathematical theory of probability in the deductive process.

3. The reader may at this point have the impression that I consider any D-N derivation to represent a good scientific explanation. This is not the case. There are well-known counterexamples to show that some derivations of the D-N form are not explanatory, or at least do not

represent causal explanations. One celebrated example utilizes the law for the period of a simple pendulum

$$\tau = 2\pi(l/g)^{1/2}, \tag{2.13}$$

where τ is the period of swing, l is the length of the pendulum, g is the acceleration of gravity, and π is the familiar geometrical constant, 3.1415 … . Now suppose that we have a pendulum with $l = 99.3$ cm, and suppose that $g = 980$ cm/sec². These facts, plus the law-sentence (2.13), presumably explain why $\tau = 2.0$ sec, although this is admittedly a rather trivial explanation. On the other hand, suppose that we are given that a pendulum has $\tau = 2.0$ sec and that $g = 980$ cm/sec². Using these facts plus (2.13) we obtain a D-N derivation of $l = 99.3$ cm. However, almost no one would want to say that this second derivation provides an explanation of why the pendulum has a length of 99.3 cm. This and other examples have been used by Bromberger (1966) to show shortcomings of the D-N form of explanatory derivations, and I believe that Hempel's reply in Hempel (1965, pp. 352–353) does not alleviate the effectiveness of the example.

It should be recalled that the law (2.13) can be derived within classical mechanics using Newton's second law of motion and Newton's law of gravity. In fact, it is elementary to derive, for small amplitudes, a harmonic law of motion for the pendulum. This law gives the exact position of the bob at any time, once the initial conditions are provided. It is then from this law that one may derive (2.13). Within the theory of classical mechanics, it is clear that the law of motion and the law of gravity are quite fundamental laws, and are used together with special boundary conditions to explain various types of specialized motions. In this context, it is clear that l and g function as independent boundary conditions which help to determine the specialized law of motion of the pendulum. Then τ is also seen to be an even more derivative dispositional characteristic of the pendulum. In other words, τ is recognized as a characteristic determined by the law of motion, the law of gravity, and the boundary conditions l and g. In fact, l can be thought of as a structural attribute of the pendulum which helps to determine the pendulum's more derivative dispositional attribute τ. In general, we can use structural attributes to help causally explain dispositional attributes, but we cannot use dispositional attributes to help causally explain structural attributes. This is a very important observation which will be developed in later chapters.

We can see then that the explanation of τ using (2.13) fits well into the general context of the theory of classical mechanics. On the other hand, I see no satisfactory way of fitting into the theory of classical mechanics the derivation of l from (2.13) plus τ and g. τ is not a suitable boundary attribute to use with the fundamental laws of classical mechanics.

In general, I think that it must be granted that some D-N derivations are not explanatory, i.e., some do not represent causal explanations. Unfortunately, there are probably no general formal conditions which could be added to those already given for D-N derivations which would guarantee that we would only obtain explanatory derivations. The above discussion indicates that whether or not a derivation is explanatory will depend partly on the general theoretical framework in which it occurs. Fortunately, in the present investigation it is not necessary to have a complete set of necessary and sufficient conditions for explanations. For my purposes it will suffice to make some rather general, and I believe reasonable, assumptions about laws and causal explanations.

4. The present study is primarily concerned with dynamic theories. The structure of such theories will be discussed in the sequel. Very generally, dynamic theories contain (among other things) law-sentences about the attributes and behavior of the various kinds of things in a specified domain of things. These law-sentences may state qualitative, quantitative, or probabilistic laws.

Now the nature of causation is a topic which has received much philosophical analysis. Most empiricists agree that causation is somehow connected with the existence of laws involving causal connections. In fact, when we think of a causal connection, we most readily think of a law-sentence having the simplest form given in Chapter 1, namely,

$$(x)(\alpha x \to \beta x). \tag{2.14}$$

If (2.14) expresses a causal connection, then, in the simplest case, we would think of instances of α causing instances of β. If α and β are attributes, we may think that the attribute α is directly causally related to the attribute β. But there is really no reason why we should restrict this interpretation to *direct* causal connections. A law-sentence of the form (2.14) may express a causal relationship in which α and β are indirectly causally connected. This *indirect* causal connection may be the consequence of other laws and intermediate processes. Thus, I will allow that a law-sentence may express an indirect or a direct causal connection.

Now if we interpret causal connections broadly, there is no reason why we should restrict our considerations to qualitative laws. Suitable quantitative laws can be interpreted as expressing causal connections between the quantities involved in the respective laws. In addition, suitable probabilistic laws can be interpreted as expressing causal connections which are not deterministic, but which instead occur at some fixed relative frequency. It is not my intention to try to defend these interpretations here; that would involve a long digression into the nature of causation. Very helpful discussions of these issues can be found in Mackie (1974, especially Chapters 6 and 9). It must, nevertheless, be remembered that not all true universal, or probabilistic, generalizations are law-sentences. Thus, one cannot automatically assume that true sentences of these forms represent causal connections. In general, I believe that our decisions concerning causal connections are theory-dependent. Consequently, outside of the context of an accepted theory it may be very difficult to decide with confidence whether or not a given true generalization represents a causal connection. Fortunately, this will not be a serious problem in the present investigation, which is almost entirely concerned with dynamic theories. It is reasonable to assume that the empirical generalizations contained in any adequate dynamic theory are law-sentences that represent causal connections. Thus, throughout this work I will assume that the universal, or probabilistic, generalizations contained in dynamic theories express either direct or indirect causal connections. A somewhat analogous assumption will be made concerning explanations in dynamic theories.

As indicated previously, not all D-N derivations represent explanations, or at least not all D-N derivations represent causal explanations. In fact, some philosophers wish to restrict causal explanations to explanations of particular events, for it is not customary to speak of the cause of a law. Nevertheless, it seems quite reasonable to consider, for example, the explanation of (2.1) given above as a causal explanation of this law. This explanation shows that this law is a causal consequence of more general Newtonian laws together with certain additional, causally relevant factors. The example of the explanation of the period of the pendulum is perhaps dubious because of its triviality. However, suppose that the period of the pendulum is explained in terms of fundamental Newtonian laws plus boundary conditions, in the manner which I have indicated above. I believe that it is quite reasonable to consider this a causal explanation of the period of the pendulum. It shows the period to be a consequence of Newtonian causal laws together with certain boundary conditions. Similarly, an explanation of (2.13) in terms of the fundamental Newtonian laws shows (2.13) to be a causal consequence of these laws together with the generic boundary conditions which define a simple pendulum. Similar remarks can be made about good D-S explanations of probabilistic laws, for example, about well-formulated D-S explanations one finds in the field of genetics. In general, then, we can consider a suitable D-N or D-S derivation of a causal law-sentence from other causal law-sentences to represent a causal explanation of the law-sentence which is derived. Henceforth, the term *explanatory derivation* will always be used to refer to a D-N or D-S derivation which represents a causal explanation.

Unfortunately, as was shown above, not all D-N or D-S derivations represent causal explanations. As in the case of causal laws, I suspect that the conditions for causal explanations are at least in part theory-dependent. Fortunately, I am almost entirely concerned with explanations within dynamic theories. As will be described in detail later, such theories contain some *fundamental* law-sentences and some *derivative* law-sentences; the latter are explained within the theory in terms of the former. It is reasonable to assume that a D-N or D-S derivation of a derivative law-sentence of an adequate dynamic theory from the fundamental law-sentences of that theory will be an explanatory derivation.

The assumptions I have made about the causal connections expressed by law-sentences and about explanatory derivations are not detailed explications of these notions. However, I believe that my assumptions arc quite reasonable assumptions about the general roles played by these notions within the contexts of adequate dynamic theories. Later, in discussions of particular types of dynamic theories, I will have more to say about further characteristics of law-sentences and of explanatory derivations.

B. IDENTITIES

1. In Chapter 1, Section C, I have introduced attributes and kinds in an informal manner, and I have indicated that they are to be distinguished from their extensions. Also law-sentences were introduced and briefly discussed in a previous section. Now the language of a scientific theory is often assumed to consist of certain logical and mathematical symbols plus nonlogical constants. As we saw, the 'logical symbols' are often

those used in set theory together with mathematical symbols definable in set theory. In a set theoretical language the nonlogical constants are usually considered to be predicate constants (including function symbols) which are interpreted as denoting sets, or sets of ordered n-tuples, of objects in the domain of the theory in question. Within such a language a universal law, of the simplest possible form, is represented by a law-sentence of the familiar form

$$(x)(\alpha x \to \beta x), \tag{2.14}$$

where α and β are predicates which just denote their extensions. It is often thought that such a set-theoretical interpretation of the predicates of a scientific language is adequate for all scientific purposes. I disagree with this view, and in this section I will argue that the predicates in law-sentences cannot always be interpreted as merely denoting sets, and that instead they should usually be interpreted as denoting kinds or attributes. An example will help to motivate the later, more general discussion.

   2. The example uses some predicates introduced earlier. Let '*sink x*' mean that the object x has the disposition to sink in H_2O in a gravitational field, let '*den(x)*' denote the density of x, and let '*gold x*' mean that x is a sample of gold. We then have the following law-sentences

$$(x)(gold\ x \to sink\ x), \tag{2.15}$$

$$(x)(sink\ x \leftrightarrow den(x) > den(H_2O)), \tag{2.16}$$

where '$>$' is the mathematical symbol for 'greater than'. It is obvious that (2.15) and (2.16) imply the law-sentence

$$(x)(gold\ x \to den(x) > den(H_2O)). \tag{2.17}$$

Now, intuitively, (2.15) and (2.17) state different laws. Furthermore, from information about the masses of their atoms and molecules, plus information about the crystalline and liquid structures of gold and H_2O, we could explain (2.17). Yet, in order to explain (2.16), and hence also (2.15), we would also use laws of the mechanics of fluids. If the explanations of two different law-sentences require obviously different sets of laws in their explanans, then the two law-sentences in question would seem to represent or state different laws. Finally, notice that (2.15) and (2.17) can each be derived from the other given the assumption of (2.16). But (2.16) represents a causal law which is explainable, and which indicates some kind of underlying causal relationship between (2.15) and (2.17). It is natural to conclude that (2.15) and (2.17) represent different laws by virtue of the causal connection between them which is represented by (2.16).

It is well known that law-likeness is not necessarily preserved under substitution of *accidentally* co-extensional predicates. Suppose that $L(P)$ is a law-sentence containing one occurrence of the predicate 'P', and suppose that 'P' and 'Q' are accidentally co-extensional. Let $L(Q)$ be the sentence resulting from the replacement of 'P' by 'Q' in $L(P)$. Then $L(Q)$ is not necessarily a law-sentence; it may be an accidentally true generalization. However, the example given above is not of this kind, for all three sentences involved are in fact law-sentences. What does this example suggest?

First notice that (2.15) and (2.17) are both very simple law-sentences having the general form of (2.14). They differ merely by the substitution of the nomologically co-extensional predicates '*sink x*' and '*den*(x) > *den*(H_2O)'. Also, (2.15) and (2.17) do not contain any modal operators such as 'necessarily', nor do they contain any propositional attitude operators such as 'it is believed that'. It might be suggested that they contain a suppressed operator, 'it is a law that'. However, even if this is supposed, it does not change the point of the example, for this example does not depend on any change of truth value or change of lawful status of the sentences involved. The point of the example is clear. We have a simple law $L(P)$ and another simple law $L(Q)$ which results by substitution of 'Q' for 'P' in $L(P)$. Moreover, this substitution is reversible, and 'P' and 'Q' are *nomologically co-extensional*. Yet $L(P)$ and $L(Q)$ are law-sentences which represent different laws. Furthermore, the fact that they represent different laws is *not* a result of the fact that (2.15) and (2.17) are not synonymous sentences. These two sentences represent different laws because they represent different causal relationships, and the fact that they represent different causal relationships is apparent from the fact that they are causally explained in significantly different ways.

It should be clear by now that there is a serious ontological problem about the nature of the entities which are related in lawful connections. Suppose that laws are simply special types of factual relations between *sets* of entities. Then (2.15) would state that it is a law that the extension of '*gold*' is contained in the extension of '*sink*', and (2.17) would state that the extension of '*gold*' is contained in the extension of '*den*(x) > *den*(H_2O)'. But these two extensional relationships would in fact be the same because of (2.16). But then, if laws are factual relations between sets of entities, (2.15) and (2.17) should state the same law, but they do not. In general, it is risky to draw conclusions about the denotations of two terms merely from peculiar results obtained by substituting one of these terms for the other. However, in the present example the sentences involved have very simple structures, and yet their roles in scientific theories and explanations are quite different. It is therefore quite plausible to conclude that scientists do not interpret the predicates in these sentences as merely denoting their set theoretical extensions.

I believe that, in the long run, decisions about the appropriate language and ontological framework for a given purpose must be based on a wide variety of theoretical and practical considerations. In various places throughout the remainder of this book I will provide various reasons favoring the realistic interpretation of kinds and attributes. I will interpret the predicates occurring in law-sentences as denoting kinds or as denoting attributes. I cannot give any conclusive reasons for doing so at this point, although a variety of reasons will emerge in later discussions. It should be noted, however, that this interpretation is helpful in understanding the present example. Interpret '*gold*' as a predicate denoting a kind of thing, and interpret the other predicates involved as denoting distinct attributes. Then (2.15) is a law-sentence which states that *gold* has the attribute *sink*, and (2.17) is a different law-sentence stating the different law that *gold* has the attribute of $den(x) > den(H_2O)$. With this interpretation there is certainly no prima facie reason to suppose that these two law-sentences state the same law, and this is as it should be.

Unless otherwise indicated, let us henceforth interpret the predicates in law-sentences as denoting kinds or attributes. Then a new problem immediately arises. It is possible that two different law-sentences, under this interpretation, might state the same law. Let us say that two law-sentences in a scientific language are *nomologically-equivalent, n-equivalent*, or *n-eq*, if and only if they represent or state the same law. What is the criterion for n-equivalence?

Achinstein says that two propositions expressing laws express the same law if they are either logically equivalent, or empirically equivalent. He considers two propositions to be empirically equivalent if they can be derived from one another with the help of additional empirical assumptions which are logically independent of each of these propositions. In Achinstein (1971, pp. 1, 16) it is even allowed that there may be formulations of a given law that are neither logically nor empirically equivalent.

Achinstein considers the concept of law, as used by practicing scientists, to be rather loose.

For reasons which will become clear shortly, logical equivalence of two law-sentences is sufficient for their n-equivalence. Moreover, Achinstein is correct in implying that logical equivalence is too narrow a criterion for n-equivalence. On the other hand, the above example, and others like it, show that his criterion is too broad. Other examples similar to the one above can be constructed by using other nomological co-extensionalities. Additional examples of such laws will be presented and discussed in later chapters in a different context.[1]

Let us look again at (2.16). Informally, we can say that the reason (2.16) states a law and is subject to a causal explanation is because '*sink x*' and '$den(x) > den(H_2O)$' are predicates which denote different attributes. The law-sentence (2.16) states that these attributes are co-extensional, and this fact is causally explainable. On the other hand, after some reflection, it would appear that an identity between designative terms is not causally explainable. An identity merely asserts that two terms denote the same object, and hence it is a sentence that is not, in principle, subject to a causal explanation. It would then also appear that identities of kinds and attributes might be involved in our sought-for criterion of n-equivalence. I will develop these rough ideas in the following.

3. Consider a language which describes kinds of objects in a specified domain of things, as well as attributes of these objects. This language consists of symbols used in set theory and mathematics plus a set $\mathcal{L}$ of nonlogical predicates and function symbols. $\mathcal{L}$ is the union of two disjoint subsets, $\mathcal{T}$ and $\mathcal{A}$, where $\mathcal{T}$ is the set of thing-predicates and $\mathcal{A}$ is the set of attribute-predicates. Recall that a thing-predicate denotes a kind of thing and an attribute-predicate denotes an attribute, which may be a property, relation, or quantity. The present discussion will be primarily concerned with attribute-predicates which denote properties or relations. Further attention will be given later to kinds and quantities.

Suppose $\mathcal{A}$ contains predicates 'W' and 'S' denoting white and spherical, respectively. Then it is possible to introduce by definition a new predicate 'P_1' by

$$(x)(P_1 x \leftrightarrow Wx \ \& \ Sx). \tag{2.18}$$

The sentence (2.18) tells us that any object x has the property P_1 if and only if x has the property white and x has the property spherical. Of course, (2.18) also implies that the extension of P_1 is the intersection of the extension of W with the extension of S.

We can also introduce the new predicate P_2 by

$$(x)(P_2 x \leftrightarrow Sx \ \& \ Wx). \tag{2.19}$$

Then it follows by simple logic from the two analytically true definitions (2.18) and (2.19) that

$$(x)(P_1 x \leftrightarrow P_2 x), \tag{2.20}$$

and (2.20) is an analytically true sentence in $\mathscr{L}$. We recall from Section A of this chapter that logical truths and definitions are analytic, and any sentence which is derivable from premises consisting only of logical truths and definitions is also analytic.

Now we need two observations. First of all, it would not make any sense to try to give a causal explanation of an analytic truth, for such a truth is the result of logical and definitional conventions and not empirical causes. In other words, we can say that analytic truths are not subject to causal explanations. In general, any sentence which is not subject to causal explanation is *noncausal*. Thus, analytic sentences are noncausal sentences. Also logical contradictions are noncausal sentences. As we will see below, there are also synthetic noncausal sentences.

In the second place, we should observe that, since '$Wx \ \& \ Sx$' and '$Sx \ \& \ Wx$' are logically equivalent for all x, they can be substituted for each other in any sentence of $\mathscr{L}$. Therefore, because of their definitions, 'P_1' and 'P_2' likewise can be substituted for each other in any sentence of $\mathscr{L}$. Also, it is intuitively quite reasonable to say that being white and spherical is the same property as being spherical and white. Therefore, we can say that 'P_1' and 'P_2' denote the same property. More generally, let α and β be any two predicates in $\mathscr{L}$, or defined in $\mathscr{L}$; then if

$$(x)(\alpha x \leftrightarrow \beta x) \tag{2.21}$$

is analytically true, then we can say that α and β denote the same kind or attribute. Of course, if α and β are two different primitive predicates of $\mathscr{L}$, and (2.21) is analytic, then (2.21) will be a definition of one of these predicates in terms of the other one.

4. I am concerned here with *thing-identities* between two thing-predicates and with *attribute-identities* between two attribute-predicates. In our ordinary language we are familiar with synthetic identities between certain proper names. For example, suppose there is a certain person

known to some by the name 'Bob' and to others by the name 'Robert'. Then the sentence 'Bob = Robert.' is true, but it is not analytically true. It is a synthetic identity, the truth of which must be justified by some kind of empirical confirmation. On the other hand, an analytic sentence is true because of logic and definitions, and so its truth is not subject to empirical justification. Finally, notice that 'Bob = Robert.' is not subject to causal explanation. Although its truth needs empirical justification, once its truth is established, we do not look for a causal explanation of *why* it is true.

We saw above that all analytic truths are noncausal. Thus, analytic truths are not subject to empirical justification nor to causal explanation. On the other hand, a synthetic identity such as 'Bob = Robert.' is subject to empirical justification, but it is not subject to causal explanation. It is an example of a synthetic noncausal sentence. I claim that, within $\mathscr{L}$, there can be synthetic identities between predicates. They too will be synthetic noncausal sentences. This is contrary to a rather long tradition which holds that attribute-identities must be analytically true; see, for example, Quine (1961, p. 157).

I have already argued that it is inadequate to interpret the predicates of $\mathscr{L}$ as merely denoting their extensions. I propose instead that we interpret these predicates as denoting kinds or attributes. Once we accept such interpretations, there is no reason why two predicates of $\mathscr{L}$ should not denote the same kind or the same attribute even though these predicates are not analytically equivalent. The problem is to distinguish identities from nomological co-extensionalities. In other words, suppose that we know a true sentence of the form of (2.21), and suppose that this sentence is not analytically true. In that case, the truth of this sentence will have been justified empirically. This sentence might represent an accidental co-extensionality, a nomological co-extensionality, or an identity between the predicates α, β. Since we are concerned with sentences which can be used in dynamic theories, we are not concerned here with accidental generalizations. As I stated in Chapter 1, the investigations reported in this book do not directly involve the accidental/nomological distinction. Therefore, I assume that it is known that our sentence is not an accidental generalization. Then how are we to decide whether this sentence is an identity between the predicates α, β, or whether it is a nomological co-extensionality? In many cases, it will appear impossible to obtain any *direct* empirical evidence which will decide this question. For this reason there have been some who have believed that the distinction is of no empirical significance. In fact, without going into details here, it seems to me that there have been many perplexities in recent literature about the distinction between nomological co-extensionalities and attribute-identities. The reader may consult Kim (1966), Putnam (1969), and Malinas (1973). Some of these perplexities will be considered in Chapter 5, in the context of my discussion of the requirements for the reduction of theories.

In spite of the perplexities concerning the distinction between identities and nomological co-extensionalities, I believe that the general nature of this distinction is rather simple when it is seen in the proper perspective. In order to gain more insight into this distinction, let us again consider (2.16). This sentence is clearly not an attribute-identity. First, it is intuitively not an identity. I do not believe that anyone would be inclined to think that the disposition to sink in H_2O in a gravitational field is the same attribute as the property of having density greater than the density of H_2O.

Second, it should be recalled that (2.16) is subject to a causal explanation, and it has in fact been explained. This explanation shows that, as a result of the laws of mechanics, the property of having density greater than the density of H_2O is causally necessary and causally sufficient for an object to have the attribute *sink*. The fact that these attributes are causally related in this manner is further reason for concluding that they are distinct attributes. There is a long philosophical tradition according to which causally related *events* must be distinct. As I have indicated above, identities do not appear to be subject to causal explanations. It can now be seen that this observation about identities is a kind of generalization of this philosophical tradition. If, contrary to fact, the predicates in (2.16) did denote the same attribute, then (2.16) would not be subject to a causal explanation.

A third observation is also in order here. Recall that (2.15) and (2.17) state different laws. Also recall that I am now interpreting the predicates in law-sentences as denoting kinds or attributes. Under this interpretation it is natural to conclude that two law-sentences will be n-equivalent if they represent equivalent relations between kinds and attributes. If, contrary to fact, the predicates in (2.16) did denote the same attribute, then (2.15) and (2.17) would be n-equivalent. But of course they are not.

These observations suggest the following about a nonaccidentally true generalization of the form of (2.21). If α and β are identical, then (2.21) will not be subject to a causal explanation, so it will be a noncausal sentence. On the other hand, if (2.21) is subject to a causal explanation, then α and β will not be identical. In this case (2.21) will be a nomological co-extensionality, i.e., a law-sentence. Furthermore, we can now see that identities are somehow connected with n-equivalence.

Let us now say that any sentence which is subject to a causal explanation is a *causal* sentence, even if we do not know of an adequate explanation. It seems to me that, in principle, any law-sentence is a causal sentence. Recall that in this work I am concerned with law-sentences in dynamic theories, and that I am assuming that such law-sentences represent either direct or indirect causal connections. Now, if certain kinds or attributes are causally connected, it seems impossible to imagine why, in principle, we should not be able to explain this connection in terms of other, more general, causal relationships. I do not claim that we will always be able to find such an explanation, but I do believe that it is always possible that such an explanation could be found. At any given time we will know some laws which we have not yet explained. Yet, it makes perfectly good sense to seek explanations for them, and eventually we may succeed in explaining them. We may even discover a law which we can never adequately explain. Nevertheless, I see no way of proving that such a law would be in principle unexplainable.

It might be objected that we could discover a causal connection so absolutely basic that it could not be explained in terms of any further causal connections. However, if we did discover such a connection, I see no way by which we could prove that it has this absolutely basic character. Moreover, it would be a mere empirical fact that it has this absolutely basic character. There is nothing in our concepts of causation and causal explanation which would require that it is unexplainable. Thus, were it not for a mere empirical fact, it might have been explained, which is to say that it is in principle subject to an explanation. Thus, I conclude that any law-sentence is a causal sentence, where 'causal' is used in the technical sense defined above, i.e., subject to a causal explanation. Consequently, any sentence of the form of (2.21) which is accepted as a law-sentence will be a causal sentence, and it will not be an identity.

Now suppose that we have a nonaccidentally true sentence of the form of (2.21) which is accepted as noncausal, i.e., not subject to a causal explanation. From the preceding discussion we can see that in this case α and β will not be causally connected. In such a case it would seem that the only reason that we could have for the lack of a causal connection is that α and β are identical. Unfortunately, there is no totally non-question-begging way to prove that they must be identical. However, I will adopt the convention that α and β are identical when (2.21) is a non-accidentally true, noncausal sentence. In the sequel it will be shown that this convention leads to some natural substitution principles. It will also turn out that whenever (2.21) is noncausal, then α and β play essentially the same roles in scientific laws, explanations, and theories. Thus, at least for all scientific purposes, α and β are indeed identical.

It should be clear that if a theory contains a sentence of the form of (2.21), it might be difficult in practice to decide its status. Obviously, we will know that we have a law-sentence if we have explained it. However, if it is not explained (either within the theory or by a reduction), then we will have to hypothesize whether it is a law-sentence or an identity. There is no general decision procedure which will tell us in particular cases which hypothesis to choose. However, in later chapters certain guidelines will become apparent. At any rate, the structure of our theory will be significantly affected by the choices which we do make. In particular, these choices will determine which law-sentences are n-equivalent and which are not.

The results of all of this discussion can be fairly easily summarized. Suppose that we have a true sentence of the form (2.21). Then it could fall into exactly one of three possible cases: (i) α and β are accidentally co-extensional, (ii) α and β are nomologically co-extensional (i.e., α and β are nomologically correlated), (iii) α and β are identical. As explained earlier, I am not here concerned with accidental generalizations, so my discussion above was concerned with cases (ii) and (iii). In this discussion I have argued that α and β are identical if and only if (2.21) is non-causal. Moreover, if (2.21) is analytic, then it must be noncausal. If (2.21) is synthetic, then it may either be noncausal or causal. Thus, α and β are nomologically co-extensional if and only if (2.21) is a synthetic causal sentence. In other words, (2.21) is a correlation law or an identity according as it is or is not subject to a causal explanation. Obviously, no correlation law is an identity, and no identity is a correlation law. Finally, the discussion indicates that identical predicates, and not nomologically correlated predicates, should satisfy some substitution principle related to n-equivalence. The next two subsections further develop this idea.

5. It would be an interesting and perhaps difficult problem to determine the class of all true noncausal sentences. Fortunately, my purposes do not

require a general solution to this problem. Moreover, we are only concerned here with the language $\mathscr{L}$ which contains thing-predicates and attribute-predicates applying to the elements of some domain of things. Chapter 3 contains a detailed description of certain types of languages of this kind. Later chapters will discuss more complex languages. Careful discussions of the various kinds of true noncausal sentences must be deferred until the various types of languages are described in more detail. For the remainder of this chapter, it is assumed that the language $\mathscr{L}$ is any language of the kind previously described, and it is understood that $\mathscr{L}$ contains no intensional contexts such as might be produced by modal operators or propositional attitude clauses. Also, as usual, I exclude from consideration accidentally true generalizations. Then it appears obvious that the set of true noncausal sentences of $\mathscr{L}$ contains at least the following: logically true sentences, analytically true sentences (which follow logically from definitions of defined terms), true thing- and attribute-identities, and logical consequences of sets of true noncausal sentences.[2]

This partial determination of the class of true noncausal sentences has certain interesting consequences. In particular, notice that '$(x)(Qx \rightarrow Qx)$' is noncausal, so if Q is identical with R, then '$(x)(Qx \rightarrow Rx)$' is noncausal, although it appears to be a law-sentence. If Q is nomologically co-extensional with R, then '$(x)(Qx \rightarrow Rx)$' is a law-sentence. Any law-sentence is in principle subject to a causal explanation, so no law-sentence can be noncausal, and no noncausal sentence can state a law. Also notice that '$(x)(y)(Uxy \leftrightarrow Vyx)$' might be synthetic noncausal. For instance, it follows from the synthetic identity of V with U^*, and the definition of U^* as the converse of U. If necessary, 'U^*' can be added to $\mathscr{L}$ as a defined term.

Turning now to n-equivalence, in Causey (in press) I have proposed the following:

NE: Let L_1, L_2 be law-sentences in $\mathscr{L}$. Then L_1 n-eq L_2 if and only if there is a set $\mathbf{N}$ of true noncausal sentences of $\mathscr{L}$ such that $L_1 \leftrightarrow L_2$ is derivable from $\mathbf{N}$.

It will be noticed that I have not provided any general characterization of the form of law-sentences. The exact form of law-sentences may vary somewhat from theory to theory. Criterion NE merely assumes that L_1, L_2 are given as law-sentences in $\mathscr{L}$. Suppose first that these two law-sentences are analytically equivalent, which means that they can be proved equivalent from a set of analytically true premises. Then each of these law-sentences can be derived from the other with the help of a set of premises which are all definitions. Thus, if one of these law-sentences is established, then the other can be established without any additional empirical information. Moreover, suppose that we have an explanatory derivation of one of these law-sentences, say L_1, from an appropriate set of premises. Then we can use these premises together with definitions to obtain a derivation of L_2. Since the first derivation is explanatory and the second is just like the first except for extra uses of logical principles and definitions, it seems obvious that the second derivation must be an explanatory derivation of L_2. Furthermore, this entire procedure is reversible; given an explanatory derivation of L_2, we can augment it with definitions to obtain an explanatory derivation of L_1. Thus, both of these law-sentences can be explained in essentially the same way. Also, if one of them is used as a premise in an explanatory derivation, then the other plus definitions can be substituted for it in this derivation.

It is helpful to contrast this situation with that of (2.15) and (2.17). In that case we have distinct laws. Intuitively, one can say that (2.15) and (2.17) are causally related by means of the intermediate law-sentence (2.16). On the other hand, if L_1, L_2 are analytically equivalent, then there is no causal relationship between them. Because of all of these considerations, we are compelled to conclude that L_1, L_2 are n-equivalent.

NE is a generalization of this idea. Suppose that there is a set **N** of true noncausal sentences such that L_1, L_2 can be derived from one another with the help of this set of premises. In this case, L_1, L_2 can be factually equivalent, but they are still not causally related because **N** is not subject to a causal explanation. Therefore, an explanatory derivation of L_1 can be transformed into an explanatory derivation of L_2 by the addition of **N** to the explanans, and vice versa. If L_1 is a premise in an explanatory derivation, then L_2 plus **N** can be substituted for L_1, and vice versa. In effect, once it is known that L_1, L_2 can be proved equivalent given **N**, then it can also be seen that these two law-sentences play essentially the same roles in explanations and theories. Intuitively, whatever causal relationship is expressed by one of these law-sentences is also expressed by the other, for these two law-sentences are equivalent given **N**, where **N** expresses no causal relationship.

All of these remarks are especially obvious in the case where **N** consists of a set of identities. For instance, the addition of an identity to the explanans of an explanatory derivation merely adds additional information about the denotations of the predicates involved. In general, since no noncausal sentence is a law-sentence, the addition of a noncausal sentence to an explanans does not add any new law-sentence, nor, indeed, any new information about causal connections.

All of these considerations amount to compelling reasons for concluding that the right-hand side of NE is sufficient for n-equivalence. On the other hand, these same considerations also make it clear that the right-hand side of NE is necessary for n-equivalence. If L_1, L_2 *cannot* be proved equivalent from a set of noncausal sentences, then L_1, L_2 will be subject to essentially different explanations and, in general, will play significantly different roles in explanations and theories. Thus, L_1, L_2 will fail to be related in ways which, intuitively, are necessary for n-equivalence. Therefore, I conclude that NE provides a necessary and sufficient condition for nomological-equivalence. This conclusion will be further strengthened by observation of the systematic import of this criterion in a number of theoretical contexts.

One obvious type of application of NE is its use in the formulation of various substitution principles satisfied by identities. In my later discussions of theory reduction, the following will be very useful: Suppose that L_1 contains occurrences of predicate α, L_2 contains occurrences of predicate β, and L_1 and L_2 are uniform substitution instances of each other under substitution of β for α in L_1, and α for β in L_2, where α and β denote the same kind or attribute. Then L_1 n-eq L_2.

Of course, in order to obtain n-equivalence by such a substitution, it is necessary that the denotations of α and β be the same and not merely nomologically co-extensional. For example, let the following law-sentences be formulated in $\mathscr{L}$:

$$(x)(Ax \rightarrow Bx), \tag{2.22}$$

$$(x)(Ax \rightarrow Cx), \tag{2.23}$$

$$(x)(Bx \leftrightarrow Cx). \tag{2.24}$$

It should now be clear that (2.22) and (2.23) state different laws because they represent different causal connections involving the different attributes B and C. If B and C were identical, then (2.22) would be equivalent to (2.23) solely by virtue of this noncausal identity of reference. However, actually (2.22) and (2.23) are more weakly equivalent by virtue of the causal law (2.24). This weak equivalence does not follow from the references of the predicates in $\mathcal{L}$, but rather from the causal law (2.24). If L_1 and L_2 are law-sentences which can be proved equivalent *only* under the substitution of nomologically co-extensional predicates, then L_1 and L_2 represent different laws.

6. Some further remarks about identities are now in order.

a. It will be noticed that most of the preceding examples involved identities of properties, although the general principles discussed apply to all thing-identities and attribute-identities. Actually, identities between predicates denoting relations should cause no special problems, for they behave essentially as do identities between predicates denoting properties. Identities between thing-predicates do cause some special problems of interpretation which cannot adequately be discussed until we have examined thing-predicates in more detail. Beginning in Chapter 3, I will have much to say about thing-predicates. Eventually it should become clear that NE does indeed apply to such predicates. I will not try to show this at this point.

b. It may not be immediately clear how to apply NE to quantities. As we saw in Chapter 1, a given quantity can be measured with different representation functions. Depending on the scale type of the quantity in question, these representation functions will be related by various types of mathematical transformation functions. However, in the usual formulations of quantitative law-sentences, particular representation functions are not denoted. Instead, a function symbol is used to denote the quantity involved, and the law-sentence is stated in such a way as to hold regardless of the particular representation function which may be substituted into it. We have previously seen examples of such formulations, and they are quite familiar in the physical sciences.

I adopt the view that a quantity is a kind of generalized property which can be quantified by means of various representation functions. The function symbols in quantitative law-sentences formulated in the customary manner are interpreted as denoting these generalized properties. For example, 'v' in (2.10) denotes a certain property of the falling body independently of the particular representation function used and regardless of the measuring procedures utilized. Thus, quantities are rather theoretical, and they are not considered to be operationally defined, although various empirical operations will be needed to determine particular methods of measurement and particular representation functions. Under this interpretation of quantity symbols, how are we to understand various mathematical substitutions of one quantity expression for another?

Consider an idealized Hooke's law spring which has zero length when no force is applied, and which in general obeys the law-sentence

$$F = -k\ell \tag{2.25}$$

where F is the applied force, ℓ is the length the spring is stretched when F is applied, and k is the spring constant which depends both on the particular spring to which the law-sentence is applied and on the units of measurement of F and ℓ. We know very well that force and length are different quantities and that they are not mutually substitutable in all law-sentences. In addition, it should be clear from (2.25) that force and length are different quantities because k depends not only on the units of measurement, but also on the spring which is used. Since force and length are measured on ratio scales, their various representation functions are related by similarity transformations. If force and length were the same quantity, a given representation function for one could differ from a representation function for the other only by a constant depending only on the units of measurement. This is not the case with k in (2.25).

Now suppose that Q_1 and Q_2 are two quantity names for quantities measured on ratio scales. Let x range over the set of all objects to which both of these quantities apply, and let f_1, f_2 range over all representation functions for Q_1, Q_2, respectively. Let a_{12} be a positive real number which is determined completely by a particular choice of representation functions. Finally, suppose that it is a synthetic truth that for every f_1, f_2, there is a_{12} such that, for every x, $f_1(x) = a_{12} f_2(x)$. This would normally be written as

$$Q_1 = aQ_2 \tag{2.26}$$

where it is understood to hold for every x and it is understood that a depends only on the units of measurement. A constant such as a, that depends only on units of measurement, is usually called a 'universal constant'. A sentence of the form (2.26), where a is a universal constant, at least suggests the possibility that the quantities involved are identical, for it suggests that a may merely indicate a scale transformation between two representation functions.

Unfortunately, quantity-identities cannot be determined merely from the presence of universal constants. Consider the law, $E = h\nu$, for the energy E of a photon of frequency v, where h is Planck's universal constant. We certainly do not identify energy with frequency on the basis of this equation. In contrast, there is the equation, $e = (3/2)kT$, where e is the mean translational kinetic energy of a molecule of an ideal gas, T is the absolute temperature of the gas, and k is here Boltzmann's universal constant. It can be argued that this equation should be interpreted as a quantity-identity. This argument depends on some aspects of theory reduction to be discussed in Chapter 5. At any rate, there does not appear to be any simple test for quantity-identities, and this should come as no surprise. However, it does appear that identical ratio scale quantities must at least satisfy an equation of the form (2.26), where a is a universal constant.

At least this discussion should show that there is no reason why NE should not apply to quantities. If Q_1 and Q_2 are accepted as identical, on the basis of (2.26) and other grounds, then Q_1 can be substituted for Q_2 in a law-sentence containing Q_2, to obtain another law-sentence n-equivalent to the first one. In addition, if Q_1 and Q_2 are identical, then (2.26) is noncausal and hence not subject to an explanation. On the other hand, if (2.26) is a causal law, then it is in principle explainable, and substitutions of Q_1 for Q_2 in law-sentences will change these law-sentences into other law-sentences representing different laws.

This discussion should clarify the status of (2.11) which was used as an explanans premise in the earlier example involving the laws of falling bodies. It will be recalled that (2.11) identifies g with the quantity $\gamma M/R^2$. The equation (2.11) does not merely assert that g and $\gamma M/R^2$ have the same numerical values in the same units of measurement. This equation asserts that g and $\gamma M/R^2$ are the same quantitative property. This equation is therefore interpreted as an attribute-identity and not as a law-sentence. However, it is a synthetic identity, and hence needs to be empirically justified. Nevertheless, since it is not a law-sentence, it is not subject to causal explanation. I believe that this interpretation of (2.11) conforms precisely with the scientific practice regarding this equation, and others like it. Of course, such an interpretation applies only to equations which express genuine identities. Numerical laws are not identities, and they must be interpreted as laws and not as identities.

c. It will be observed that the examples and discussions above were directly concerned with universal laws. However, NE is intended to apply as well to statistical laws. The mathematics of probability theory is based on set theory, so the calculation of probability values is based on the extensions of predicates. But this does not imply that we must determine n-equivalence of statistical law-sentences merely on an extensional basis. Using nomological co-extensionalities, it is possible to construct examples of statistical law-sentences which are extensionally equivalent, but not n-equivalent.

d. In our ordinary language, and in some formal systems of logic, use is made of definite descriptions such as 'the one and only person in room r at time t', where r is some specified room and t is some specified time. Sentences which I call '*quasi-identities*' are sometimes stated using definite descriptions, e.g., 'John = the one and only person in room r at time t.' It might be suggested that such sentences are examples of identities which are subject to causal explanation.[3] At first glance, it seems to make some sense to ask, 'Why is it the case that John = the one and only person in room r at time t?' However, it can easily be seen that an adequate answer to this question does not consist of a causal explanation of an identity statement. In order to answer this question we must provide explanations for two explananda: (i) 'John is in room r at time t.' and (ii) 'For any person y, if $y \neq$ John, then y is not in room r at time t.' It is obvious that neither one of these explananda is an identity sentence, i.e., a sentence stating the identity of reference of two designating terms. Moreover, if we instantiate (ii) by some person a other than John, then we would have to explain why a is not in room r at time t.

It can be questioned whether quasi-identities are genuine identities, i.e., sentences merely stating the identity of reference of two terms. But, even if they are interpreted as identities, this example shows that they are not explainable *as identities*. An alleged explanation of a quasi-identity really consists of two separate explanations. It does not consist of an explanatory derivation of the quasi-identity sentence, nor of any other identity sentence.

It will be recalled that $\mathscr{L}$ does not contain any proper names, so no quasi-identity of the type just considered can be formulated in $\mathscr{L}$. However, not all quasi-identities refer to particular objects. Some may refer to kinds of things and some may refer to attributes. For instance, consider 'Diamond = the hardest material.' If we suppose that this is indeed true, then its alleged explanation would in fact require explaining why: (i) all samples of diamond have some particular degree of hardness, and (ii) any sample of any other material has hardness less than that of diamond. It is quite clear that these explanations are not explanations of identities. It should also be noticed that the scientific import of

'Diamond = the hardest material.' can be more clearly exhibited by avoiding the use of this sentence entirely, and by using instead (i) and (ii).

A somewhat similar example involving attributes is 'Yellow = the color of gold.' Actually, this is a very peculiar sentence which may be subject to various interpretations. Nevertheless, after a little consideration it soon becomes clear that the only sensible way to "explain" this sentence is to explain why all samples of gold are yellow. In fact, this alleged identity appears to be a disguised universal generalization. At any rate, the scientific import of this quasi-identity can best be exhibited by avoiding its use, and by using instead the universal generalization.

It is sometimes suggested that scientists need to use definite descriptions of kinds or attributes in order to refer to theoretical entities. Possible forms of such descriptions are: 'the substance which has such and such characteristics' and 'the property which causes such and such effects'. I believe, on the contrary, that descriptions with forms of this type have no place in scientific theories. Suppose that one wishes to hypothesize that there is some heretofore unknown property which causes certain effects. Then he should introduce a new attribute-predicate into $\mathscr{L}$. He should then hypothesize appropriate theoretical law-sentences containing occurrences of this attribute-predicate. These law-sentences should then be subjected to empirical tests, and they should be used to help provide explanations of the certain effects in question. There is then no need for the use of the question-begging description, 'the property which causes such and such effects'.

Fortunately most, if not all, scientific theories can be formulated in languages which are sufficiently concise to preclude the formation of the rather bizarre types of definite descriptions considered above, so quasi-identities are no serious concern. However, it should be remembered that most quasi-identities are causal sentences because they are subject to the indirect explanations described above. But these indirect explanations are not explanations of identity-sentences, so quasi-identities are not counter-examples to my thesis that identities are noncausal.[4]

C. THEORIES

The results of Sections A and B can now be combined to produce a final statement of the general form of D-N derivations. In general such a derivation may have four kinds of premises: law-sentences, boundary

conditions, analytic sentences, and synthetic noncausal sentence
course, conditions (i) through (vi) of Section A, Subsection 1, are imp
as previously. I also continue to consider D-S derivations as sp
cases of D-N derivations.

Needless to say, the resulting characterization of D-N derivations
is quite general. Fortunately, in most of the work of this book, I will
only need to consider D-N derivations which utilize three kinds of
premises, namely, law-sentences, identities (which may be either analytic
or synthetic), and, possibly, other analytic sentences. Moreover, the D-N
derivations I will consider will usually have explananda which are law-
sentences. These restrictions result from the type of theories I will
consider. It will be convenient now to consider just a few of the most
general features of the structure of these theories. These features may not
apply to all theories, but they are possessed by a large class of theories.
A few exceptions will be discussed in Chapter 8.

The kind of theory which will largely concern us can be formulated
in a language $\mathscr{L}$ of the type I have been discussing. The theory consists
of a set of sentences $\mathbf{T} = \mathbf{F} \cup \mathbf{I} \cup \mathbf{D}$, where $\mathbf{F}$ is the set of *fundamental*
law-sentences of $\mathbf{T}$, $\mathbf{I}$ is a set of true identities, and $\mathbf{D}$ is the set of
derivative law-sentences of $\mathbf{T}$. Explanations in $\mathbf{T}$ are represented by suitable
D-N derivations from explanans which may contain fundamental law-
sentences, identities, and analytic sentences. Each law-sentence in $\mathbf{D}$ is
explainable in this manner from sub-sets of $\mathbf{F} \cup \mathbf{I}$. No law-sentence in $\mathbf{F}$
is so explainable within $\mathbf{T}$, although it is possible that the law-sentences in
$\mathbf{F}$ might be explainable from the fundamental law-sentences of some other
theory.

Suppose that $\mathbf{E}$ is the explanans of an explanatory derivation of
derivative law L_1 and suppose that L_1 n-eq L_2. Then there is a set $\mathbf{N}$ of
noncausal sentences from which $L_1 \leftrightarrow L_2$ is derivable. If $\mathbf{N} \subseteq \mathbf{T}$, then L_2
is derivable in $\mathbf{T}$ from $\mathbf{E} \cup \mathbf{N}$. In general, this new derivation should be
considered to represent a different explanation because it will usually
involve extra premises which are synthetic identities. However, on the basis
of the analysis in Section B, we know that it is only different in a very
weak sense because both derivations use the same set of causal premises,
i.e., the same set of law-sentences, in order to explain the same law.[5]

If two explanatory derivations represent the same explanation they
are said to be *explanatorily-equivalent, e-equivalent*, or *e-eq*. It is difficult
to give a complete criterion for e-equivalence; however, the following
sufficient condition from Causey (1972a) is a plausible consequence of NE:

EE: Let α, β be predicates, and let $\mathbf{D}_1$, $\mathbf{D}_2$ be explanatory derivations such that $\mathbf{D}_2$ is obtainable from $\mathbf{D}_1$ by uniform substitution of β for α, and such that $\mathbf{D}_1$ is obtainable from $\mathbf{D}_2$ by uniform substitution of α for β. Then, if α and β denote the same kind or attribute, then $\mathbf{D}_1$ e-eq $\mathbf{D}_2$.

Law-sentences and explanatory derivations are linguistic representations of certain types of causal relationships. NE and EE imply that *suitable* substitutions of identicals leave invariant the causal relationships that are represented. This is not surprising since identities are noncausal sentences. Nevertheless, the requirement of mutual uniform substitution in EE is very important. Since $\mathbf{D}_1$, $\mathbf{D}_2$ are mutual uniform substitution instances of each other, $\mathbf{D}_1$ contains occurrences of α but no occurrences of β, and $\mathbf{D}_2$ contains occurrences of β but no occurrences of α. In particular, neither derivation can contain occurrences of the identity '$\alpha = \beta$'. EE is therefore somewhat limited in application. In spite of this limitation, EE does have important applications as will be seen in Chapter 5.[6]

Beginning in the next chapter, much more will be said about the structure of special kinds of theories. For the moment we should keep in mind the general roles of $\mathscr{L}$, *Dom*, $\mathbf{F}$, $\mathbf{I}$, and $\mathbf{D}$. In addition, I will sometimes refer to the *ontology* of $\mathbf{T}$; this ontology will include all of the various kinds of things in *Dom*, as well as all of the various attributes referred to within the theory. At this point the reader may feel that I have not adequately demonstrated an essential role for identities in theories. Actually, the need for identities will become most apparent when I discuss the conditions for the reduction of one theory to another. Later, when I discuss this topic of reduction, I will again discuss identities in some detail. Until that point is reached, I will, when it is appropriate, discuss theories using set-theoretical interpretations of the predicates. In this manner I believe that I can clearly show the limits of applicability of set-theoretical interpretations of theoretical languages.

NOTES

[1] Additional examples of nomological co-extensionalities can also be found in Bergmann (1966), where they are called "cross-sectional laws".

[2] I find the notion of an accidentally true generalization to be quite obscure. At any rate, I leave it as an open problem to explicate this notion and to determine whether such sentences may be causal or noncausal or either. Since this is left as an open problem, it is very important to remember that all of the principles and arguments presented in

this book are concerned with nonaccidentally true sentences unless specifically stated otherwise.

[3] For a helpful discussion of this, and related issues, see Kim (1969). Although stated in somewhat different terms, Kim's position on this issue appears to be generally compatible with mine.

[4] Parts of Section B are based on Causey (in press), which in part contains developments of some ideas in Causey (1972a). I wish to thank many persons, and in particular C. A. Anderson, for useful comments on this earlier work.

[5] Nickles (1971) discusses the use of what he calls "identificatory premises" in the explanans of explanatory derivations. Unfortunately, he is not clear about the distinction between genuine identities and nomological co-extensionalities. These must be carefully distinguished. The analysis in Section B shows that adding a nomological co-extensionality to an explanans can significantly change the causal relationships expressed in the explanation and also result in a changed explanandum.

[6] It should be immediately obvious that EE does *not* apply to a trivial derivation of a law $L(\beta)$ from premises $L(\alpha)$ and '$\alpha = \beta$'. Moreover, such a trivial derivation could not be explanatory (i.e., a causal explanation) because $L(\alpha)$ n-eq $L(\beta)$.

EE is a carefully formulated and rather subtle condition, and it must be used with care. Enç (1976, pp. 288–289) apparently attempts to provide a counterexample to EE, but his example does not satisfy the antecedent of EE. The interested reader may also wish to compare EE with Achinstein's (A) and the discussion of (A) in Achinstein (1974, pp. 258–260).

CHAPTER 3

THEORIES WITH STRUCTURED WHOLES

Chapters 3, 4, and 5 are concerned with the conditions for the reduction of one theory to another. In particular, I will focus attention on the detailed conditions for microreductions. In order to do this, Chapter 3 will first contain an examination of the logical and semantical structure of theories with structured wholes. Such theories have in their domains some elements which are considered simple and indecomposable, and other elements which are structured wholes whose parts are various kinds of the indecomposable elements.

Section A of the present chapter is a brief introduction to the aims of Chapters 3, 4, and 5. In Section B I discuss an example from chemistry which will help to motivate the general conditions I will impose on theories with structured wholes. In Section C I will discuss the kind of language needed for the formulation of such theories. Section D describes some important logical and empirical conditions satisfied by structured wholes. Finally, in Section E, I discuss some general conditions satisfied by the laws of theories with structured wholes.

A. Introduction

Occasionally the laws of one scientific theory T_2 are explained by another theory T_1. When such an explanation satisfies certain conditions, it is usually said to provide a reduction of T_2 to T_1. Chapters 3, 4, and 5 are concerned with certain types of reductions.

I use the term 'theory' here in the sense described in Chapter 2, Section C. I will assume that T_1 and T_2 are concerned with the objects in the domains, Dom_1 and Dom_2, respectively. Now it often happens in reductions that the objects in Dom_2 are assumed to be, or discovered to be, wholes whose parts are elements of Dom_1. In such a case the reduction is usually called a 'microreduction'. However, in this book I will use this term in a slightly different sense.

The basic necessary condition for a *microreduction* of T_2 to T_1 is that it be a reduction of T_2 to T_1 in which the elements of Dom_2

are identified with certain elements in Dom_1. It is further assumed that Dom_1 always contains a subset of *basic elements* which, from the point of view of T_1, are not structured wholes. In addition, Dom_1 may contain *compound elements*, which are structured wholes whose parts are basic elements. Thus, depending on the particular microreduction and the particular element, an element in Dom_2 may be identified with either a basic or with a compound element of Dom_1. Therefore, my notion of microreduction includes the usual notion as a special case.

As I indicated earlier, basic research is concerned with increasing our knowledge of the world, and also with increasing our understanding of the world. Successful microreductions can produce large increases in understanding by showing that T_2 is understandable in terms of T_1. Some actual or possible microreductions are: the reduction of thermodynamics to statistical mechanics, the reduction of the wave theory of light to the electromagnetic theory, the reduction of chemistry to the atomic-molecular theory, the reduction of biology to chemistry and physics, and the reduction of sociology to individual psychology. Other, more specific examples can be given, and will be mentioned in later sections.

Another indication of the great importance of microreductions is the fact that they are frequently involved in programs for the unification of science. One such program is the one of Oppenheim and Putnam (1958). They assume that the objects in the world are arranged in a hierarchical set of levels. It is assumed that the objects of any level, with the possible exception of a set of indecomposable objects on some lowest level, are composed of parts from lower levels. The program for the unification of science then consists of the attempt to microreduce successively the theories of all levels, except the lowest, down to the theory of the lowest level. Such a program is fraught with many difficulties, see Causey (1968a), but nevertheless it has already been accomplished to some extent. Furthermore, it seems to be an important guiding principle for present and future scientific research.

Needless to say, much has been written about the nature of reduction. Until recently, most of this literature has discussed primarily formal problems connected with derivations of T_2 from T_1. However, reductions involve very special types of, usually approximative, derivations. Especially in the case of microreductions, the theories, their languages, and the reductive derivations must satisfy many detailed conditions. Moreover, most of these conditions are of considerable philosophical significance. For instance, it would be useful to have a fairly detailed characterization of the role of structures in microreductions. It would also be useful to have a characterization of the general syntactical and semantical features of the sentences which are used in establishing a derivational connection between T_1 and T_2. It is also desirable to know exactly what is identified in microreductions and how these identifications are expressed. Failure to attend to these kinds of details can lead to confusions and to overly simple arguments for or against the possibility of certain reductions.

In addition to its theoretical shortcomings, most of the literature on reduction also has had the disadvantage of *only* providing conditions to be met by *completed* microreductions. Such work is useful, but it should be supplemented by discussions of some of the *practical* difficulties faced by scientists when they are actually attempting to construct microreductions. In particular, when scientists are attempting a reduction they will usually modify both their theoretical languages and their theoretical assumptions in the process. It would be methodologically useful to have at least some rules of thumb for guiding these modifications and keeping them within reasonable limits. The methodological significance of schemas for completed microreductions would then become more apparent.

In Chapters 3, 4, and 5, I will develop a detailed characterization of microreductions which will throw light on all of the above issues, as well as several others. These chapters are based on a combination, and further development, of work presented in Causey (1972a) and Causey (1972b). It turns out that in order to characterize microreductions in a scientifically realistic manner, it is necessary to make some fairly strong assumptions. The characterization will therefore consist of much more than a bare syntactical skeleton. In particular, it will involve certain homogeneity assumptions, and therefore it will be a schema for *uniform microreductions*. Nevertheless, in spite of this limitation of scope, I believe that it corresponds to most actual scientific reductions much more closely than any previous model of reduction. In order to try to justify this last claim, the next section will discuss a few aspects of the reduction of chemistry to the atomic-molecular theory. This example from chemistry is presented especially in order to motivate my later introduction of certain homogeneity conditions. The aspects of the reduction of chemistry which are emphasized in the example are therefore those aspects which make it a uniform microreduction.

B. AN EXAMPLE FROM CHEMISTRY

Very roughly speaking, chemistry is the science of matter. But a sample of matter is usually considered to be either homogeneous or heterogeneous. Again speaking roughly, a *homogeneous* sample of matter is a sample which has the same attributes throughout the sample. A *heterogeneous* sample is one which is not homogeneous. However, the situation is unfortunately not this simple, for this is a weak sense of homogeneity which applies only to particular samples. But the chemist is not primarily interested in particular samples. He is more concerned with classifying *kinds* of matter. As will become clear shortly, this requires considering a stronger criterion of homogeneity for the purposes of classification. There are also other complications to be considered.

In the first place, the chemist assumes by convention that only a certain class of attributes is relevant to classification. For instance, a certain diamond may have the property of having been mined in South Africa. However, to the chemist, such an attribute is 'accidental' and not relevant for the purpose of classifying the kind of matter which it is. Similarly, the mass of a sample is not used for classification; a piece of gold just forms two pieces of the same kind of matter (gold) when it is cut in half. In view of this, in order to decide whether or not a sample of matter is of a certain kind, one must consider only certain relevant *classifying attributes.*

However, there are other complications. Matter can exist in different states. Ordinary matter (i.e., not plasmas) can be either solid, liquid, or gaseous. Thus, a sample of pure water (H_2O) can be either ice, liquid water, or water vapor. The water sample clearly has different attributes in these different states. However, for the purposes of chemical classification, these differences are ignored by defining attributes in a more general way as a function of the state. For instance, one can say that water has such and such property (e.g., density = 0.917 g/cm^3) under such and such conditions (when it is solid at 0° C. and under one atmosphere of pressure). Thus, one can proceed with the problem of classifying matter after he has made proper allowances for the existence of different states.

In the chemist's classification, a kind of matter is a *substance,* and a substance name (e.g., 'gold') is a mass term. I adopt the convention that the extension of a mass term at any time is the set of all of its possible samples at that time. In classifying substances, a chemist requires a substance to be homogeneous in the *strong sense* that *all samples* of it have the same classifying attributes. Of course, as we have just seen, this implies, in particular, that any two (pure) samples of a substance which are in the same state have the same attributes. Moreover, the chemist is not only interested in physical attributes, but also in chemical attributes. Thus, he is interested in the kinds of transformations and reactions a substance can undergo which will change it, possibly together with other substances, into still further substances. When such chemical properties are studied, it is found that some substances can be decomposed into sets of other substances, whereas others are indecomposable. The former are *compounds*, and the latter are *elements*. Moreover, at least in classical (Nineteenth Century) chemistry, the composition by weight of a compound is considered a chemical property relevant to classification. Therefore, a substance which is a compound must be such that all samples of it have the same composition of the elements which compose it.

There is a further complication which must be mentioned. It is often said that a sample of a *solution* (which is a uniform mixture of more than one substance) is homogeneous in the weak sense that it has the same properties throughout. However, solutions, such as sugar dissolved in water, do not have definite compositions. Their compositions are determined by the accident of their formation, and these compositions· can vary, for they are not determined to be exact numbers by chemical laws. Thus, solutions are not substances, since they are not really homogeneous when we require all samples to have the same composition. This conclusion applies even to solid solutions, such as alloys. For example, the alloy bronze is a solid solution of the element copper and the element tin. But bronze is not a substance (not a compound) because various solid solutions of copper and tin with different compositions are all called 'bronze'.

Now we need to consider substances in more detail. As we have seen, the names of substances (e.g., 'gold', 'carbon', 'sodium chloride') are mass terms. In addition, these names are also thing-predicates, for a substance-name names a class of things (the set of all of its samples) rather than an attribute. But the extension of a substance name is such that all of its members have the same classifying attributes. For this reason I say that this extension is a *homogeneous set* and that the substance name is a *homogeneous thing-predicate*.

Classical macroscopic chemistry is concerned with classifying the various kinds of substances and with studying the laws which describe their attributes and their transformations (reactions). The atomic-molecular theory was invented in order to provide a microreductive explanation of these substances and the laws governing them. Unfortunately, this is a complicated procedure because a macroscopic sample of a substance consists not merely of a collection of atoms or molecules, but rather of a structured collection of atoms or molecules. Thus, the reductive explanation involves not only molecular structure, but also structures whose parts are molecules. Thus, it is at least a two-level microreduction. However, fortunately, it is an empirical fact that samples of substances cannot be infinitely divided, for there are empirically smallest samples of each substance. For the sake of our example, I will consider classical chemistry to be concerned only with *empirically smallest samples* (ESS) of substances. This is a simplification which was often tacitly made in Nineteenth Century chemistry in order to avoid dealing with structures whose parts are molecules.

Thus we consider a theory T_s which is about a domain of things (ESS). This theory uses a set of primitive nonlogical predicates $\mathscr{L}_s$. Some of these predicates are thing-predicates (e.g., 'gold', 'sodium chloride') and some are attribute-predicates (e.g., 'boiling point $= 100\,^{\circ}C$', 'dissolves in HCl solution'). Conveniently, the thing-predicates of $\mathscr{L}_s$ are not mass terms because T_s is about ESS. Yet these primitive thing-predicates are all homogeneous thing-predicates. Finally, T_s is a set of laws about the ESS and their attributes.

The object of the reduction is to reduce T_s to T_{am}, the atomic-molecular theory. T_{am} is a theory which uses a set $\mathscr{L}_{am}$ of primitive nonlogical predicates. The thing-predicates in $\mathscr{L}_{am}$ name kinds of atoms (e.g., 'hydrogen atom', 'gold atom'). These are also homogeneous thing-predicates because all atoms of the same kind have exactly the same atomic classifying attributes. It is also the case that any kind of atom in the set of all atoms is named by one of these primitive homogeneous thing-predicates or by some homogeneous thing-predicate definable in $\mathscr{L}_{am}$.

Of course, the atoms can form molecules. The names of molecules are defined in $\mathscr{L}_{am}$ by describing the structures of these molecules. The theory T_{am} is therefore basically a theory about atoms, but it is derivatively a theory about both atoms and molecules. The microreduction proceeds by identifying a given kind of empirically smallest sample with either a kind of atom or with a kind of molecule (e.g., 'H_2O molecule'). Needless to say, the reduction also involves much more, for the laws of

$\mathbf{T}_s$ must be explained by the laws of $\mathbf{T}_{am}$ with the help of a suitable set of sentences which make possible the appropriate explanatory derivations. These *connecting sentences* will be discussed in Chapters 4 and 5. At present, we are not concerned with them.

The following crucial points should be observed. First, $\mathbf{T}_{am}$ deals with two kinds of individuals: (i) the atoms, which are *basic* individuals, from the point of view of $\mathbf{T}_{am}$, because they are not structures composed of parts, and (ii) the molecules, which are *compound* individuals because they are structures composed of two or more atoms. Furthermore, $\mathbf{T}_{am}$ is such that a predicate naming a kind of atom or molecule is a *homogeneous* thing-predicate, and, likewise, in $\mathscr{L}_s$ a predicate naming a kind of empirically smallest sample is a homogeneous thing-predicate. Thus, when a certain kind of empirically smallest sample is identified with a certain kind of atom or molecule, the homogeneity of the predicate naming this empirically smallest sample is understandable because of the homogeneity of the predicate naming the appropriate kind of atom or molecule.

Eventually I will describe in more detail a general schema for *uniform microreductions*, which are microreductions which satisfy certain homogeneity conditions which will be defined later. I believe that most actual reductions which have been performed are uniform microreductions or nearly uniform microreductions. Hopefully this will become clear later. For the present, this example from chemistry should at least help to motivate the following general characterization of theories with structured wholes.

C. The Languages of the Theories

The primary purpose of this section is to describe certain special conditions satisfied by the language of a theory $\mathbf{T}_1$ which is a theory with structured wholes. However, since we will eventually be concerned with conditions for the microreduction of another theory $\mathbf{T}_2$ to $\mathbf{T}_1$, it will also be convenient to describe the language of $\mathbf{T}_2$ in this section. This procedure will be followed throughout the remainder of this chapter. Although this chapter is primarily concerned with $\mathbf{T}_1$, $\mathbf{T}_2$ will also be discussed whenever it is convenient to do so.

It is assumed that both $\mathbf{T}_1$ and $\mathbf{T}_2$ satisfy the general conditions for theories described in Chapter 2, Section C. In particular, $\mathbf{T}_i$ $(i = 1, 2)$ is formulated with the help of a set of nonlogical predicates $\mathscr{L}_i$. It is not assumed that the predicates in $\mathscr{L}_i$ are logically primitive; thus, some of the predicates in $\mathscr{L}_i$ may be definable in terms of others. More will be said about such definitions below.

The two theories are assumed to be deductively closed, i.e., if a sentence is derivable from a set of sentences in one of the theories, then that sentence is also in the theory. This assumption is not quite in accord with ordinary scientific usage, but it is a useful technical device. It does not cause significant loss of generality for our purposes.

Each $\mathscr{L}_i = \mathscr{T}_i \cup \mathscr{A}_i$ $(i = 1, 2)$, where $\mathscr{T}_i$ is the set of thing-predicates, and $\mathscr{A}_i$ is the set of attribute-predicates. In a microreduction we will identify the kinds of things in Dom_2 with kinds of things in Dom_1. I will eventually argue that we must also identify the attributes of $\mathbf{T}_2$ with attributes of $\mathbf{T}_1$. Thus, we will face some rather delicate problems about thing-identities and attribute-identities. In order to develop these problems carefully, I will proceed as far as is possible within a set theoretical interpretation of the languages of the theories.

As indicated previously, a basic element of Dom_1 is an element of that domain which is not, from the point of view of $\mathbf{T}_1$, a structured whole whose parts are in Dom_1. Let Bas_1 be the subset of basic elements of Dom_1. In addition, Dom_1 will usually contain some *compound elements*. These compound elements are structured wholes containing two or more parts which are elements of Bas_1. The subset of Dom_1 consisting of all of these compound elements is $Comp_1$.

In the microreduction the elements of Dom_2 are the individuals to be microreduced to Dom_1. It is not especially important whether or not elements of Dom_2 can be parts in larger structures. Therefore, we can assume that, from the point of view of $\mathbf{T}_2$, all of the elements of Dom_2 are basic elements, although, as is the case with the ESS, some of them may be transformable into others.

On the other hand, $\mathbf{T}_1$ is a theory with structured wholes. We want it to be a theory which is essentially about the basic elements of Dom_1, such that the laws about the compound elements are only derivative. More generally, we want the compound elements to be completely understandable in terms of the basic elements. In order to guarantee that this is the case, certain special conditions must be satisfied by $\mathscr{L}_1$ and $\mathbf{T}_1$. I will first discuss the conditions on $\mathscr{L}_1$.

First of all, it is assumed that there is a subset of the predicates in $\mathscr{T}_1$ which are the *basic* thing-predicates of $\mathscr{L}_1$, and which denote various kinds of elements in Bas_1. It is further assumed that no basic thing-predicate is definable in terms of other basic thing-predicates, with the possible exception of a definition which is an analytic co-extensionality between two logically primitive predicates. Finally, it is assumed that each kind of structured whole in $Comp_1$ is denoted by some thing-predicate which is defined in $\mathscr{L}_1$ in terms of basic thing-predicates together with appropriate attribute-predicates. The nature of these definitions will be discussed below.

Now, of course, the extensions of all predicates in $\mathscr{L}_2$ are subsets of Dom_2, and the extensions of all predicates in $\mathscr{L}_1$ are subsets of Dom_1. However, special attention must be given to the extensions of the predicates in $\mathscr{A}_1$. Consider an attribute-predicate, say 'A', in $\mathscr{A}_1$. 'A' might denote an attribute such as *has mass* which is possessed by both basic and compound elements. 'A' might denote an attribute such as *has atomic number 1*, which is possessed only by basic elements. 'A' might denote an attribute such as *diatomic* which is possessed only by compound elements.

Now recall that $\mathbf{T}_1$ is to be a theory which explains the structured wholes in terms of their parts, which are certain basic elements. Therefore, I will eventually require that the fundamental laws of $\mathbf{T}_1$ be formulated using attribute-predicates which denote attributes which are possessed by basic elements. It follows that there should be no attribute possessed only by compound elements which is denoted only by logically primitive predicates in $\mathscr{A}_1$. Therefore, I state the following conditions.

There is a subset of the predicates of $\mathscr{A}_1$ which are *basic attribute-predicates* and which denote *basic attributes*. No basic attribute-predicate is definable only in terms of other basic thing- or attribute-predicates, with the possible exception of a definition which is an analytic co-extensionality between two primitive predicates. Each component of a basic attribute-predicate applies to some basic element of Dom_1. Any attribute which is possessed in any one of its components only by compound elements is a *compound attribute*. Any attribute which is not basic, in particular any compound attribute, is denoted by some attribute-predicate which is defined in $\mathscr{L}_1$ in terms of basic attribute-predicates possibly together with appropriate basic thing-predicates.

Needless to say, the situation is simpler in the case of $\mathscr{L}_2$. Dom_2 has no compound elements, and so $\mathbf{T}_2$ does not involve any compound attributes. However, we do need to consider the semantics of thing-predicates in more detail, and it will be easier to begin with the simpler case of $\mathscr{T}_2$. In particular, we must consider the classification system applied to Dom_2.

For the purpose of classifying Dom_2, some of the attributes named in $\mathscr{A}_2$ may be very useful, and others considered irrelevant for the purposes of classification. As we have seen, in classifying chemical substances, there are certain so-called physical and chemical properties which are considered relevant, and other features which are considered irrelevant. Examples of relevant attributes might be color, hardness, melting point, reactivity with oxygen, ability to form hydrides, etc. Examples of irrelevant attributes are date of discovery, mass of a particular sample, position of the largest known sample, and (at least nowadays) toxicity to humans. I will assume, in an analogous way, that there is a subset of $\mathscr{A}_2$ the predicates of which specify the set of relevant attributes for classifying the elements of Dom_2. This is the subset of *classifying attributes* of $\mathscr{A}_2$. Using these attributes, one can characterize homogeneity. Let S be any subset of Dom_2. Then S is *homogeneous* if and only if any two elements of S have exactly the same classifying attributes. If S is not homogeneous, then it is *heterogeneous*.

It is important to understand that I am not assuming that there are any absolute criteria for deciding which attributes are classifying attributes, and which are not. On the contrary, the choice of classifying attributes is relative to the theory under consideration; in this case, T_2. I am only assuming that the theory does distinguish classifying from nonclassifying attributes, and that it is in terms of this distinction that the theory distinguishes kinds of things. This will become clear shortly.

Let P be a predicate in, or defined in, $\mathscr{L}_2$. Then P is a *homogeneous predicate* iff Ext_P is a homogeneous set, where 'iff' is the familiar abbreviation for 'if and only if'. One very important condition we impose on $\mathscr{T}_2$ is that the logically primitive predicates in $\mathscr{T}_2$ name only homogeneous subsets of Dom_2 and that any homogeneous subset of Dom_2 is named by some logically primitive thing-predicate in $\mathscr{L}_2$. In a similar way, it will also be assumed that there are homogeneous thing-predicates in $\mathscr{L}_1$. More will be said about these predicates later.

Fortunately, there is a simple way to determine the homogeneous subsets of Dom_2. For any x, y, in Dom_2 we have $E_2 xy$ iff x and y have exactly the same classifying attributes. It is easy to see that E_2 is an equivalence s-relation, i.e., it is reflexive, symmetric, and transitive on Dom_2. E_2 is the *natural equivalence relation* on Dom_2, and, since it is an equivalence s-relation, it partitions Dom_2 into disjoint *natural equivalence classes*. Clearly, each of these equivalence classes is a homogeneous subset of Dom_2, and each homogeneous subset of Dom_2 is one of these equivalence classes.

It is now easier to grasp the character of $\mathcal{T}_2$. Each logically primitive predicate in $\mathcal{T}_2$ is a homogeneous thing-predicate which names a natural equivalence class of Dom_2. Moreover, each natural equivalence class of Dom_2 is named by some homogeneous logically primitive thing-predicate in $\mathcal{T}_2$.

In a similar way, it is assumed that a subset of the attribute-predicates in $\mathcal{A}_1$ is the set of classifying attribute-predicates for Bas_1. Thus, one defines E_1, the natural equivalence s-relation on Bas_1, and Bas_1 is thereby partitioned into disjoint natural equivalence classes. It is assumed that each basic thing-predicate in $\mathcal{T}_1$ names a natural equivalence class of Bas_1, and that each such natural equivalence class is named by some basic thing-predicate in $\mathcal{T}_1$. Finally, it is assumed, that the only (logically) primitive predicates in $\mathcal{T}_1$ are primitive basic thing-predicates. Thus, any kind of compound element is denoted by a defined thing-predicate.

It is possible that the theories might make use of some heterogeneous thing-predicates. It should be observed that the above conditions require that any heterogeneous thing-predicate be a defined thing-predicate. There are various ways of formulating such definitions, for example, *bronze is a solid solution of copper and tin.*

It is very important that neither $\mathcal{T}_1$ nor $\mathcal{T}_2$ have primitive heterogeneous thing-predicates. This fact will play an essential role in my subsequent conditions for reduction. Indeed, it is now possible to say that a *uniform microreduction* is a microreduction such that: (i) $\mathcal{T}_1$ and $\mathcal{T}_2$ contain no primitive heterogeneous thing-predicates, and (ii) the *Uniformity Condition* is satisfied. This condition will be stated in the next section.

D. Structures and homogeneity

In this section I will describe Dom_1 in some detail. I am particularly concerned to show how compound elements can become involved in T_1 and Dom_1. In order to avoid confusion, I adopt the convention that compound elements have *at least two parts* which are basic elements. A compound element is also required to be a structured whole of a *finite* number of basic elements.

In this discussion the concept of structure will intentionally be left rather broad. Since there can be a multitude of different kinds of structures in different theories, this broadness is necessary in order to obtain a very general characterization of theories with structured wholes.

In the first place, in order to describe a kind of structure, one must specify a finite set of two or more basic elements and also specify some structural relationship which holds between these elements. It will normally be required that this structure be reasonably stable under some specifiable conditions. The criterion for 'reasonable stability' will naturally depend on the theory at hand.

In order that $\mathscr{L}_1$ be able to describe structures, it is necessary for $\mathscr{A}_1$ to have some *structural relation-predicates*, or that such predicates be definable in $\mathscr{A}_1$. It is in terms of these structural relation-predicates, which I will usually call '*structure-predicates*', that the various kinds of possible structures of compound elements are described. For example, in classical chemistry, molecular structures are described in terms of chemical bonds and valences, of which there are several types.

Now recall that Bas_1 is partitioned by E_1 into disjoint natural equivalence classes of basic elements. We say that two basic elements are of the same *kind* (or *type*) iff they are in the same natural equivalence class. Since these natural equivalence classes are named by basic homogeneous thing-predicates of $\mathscr{L}_1$, these basic homogeneous thing-predicates name the various *kinds* of basic elements. There will also be various kinds of structures, about which more will be said shortly. At the present it should be noticed that all of the various kinds of structures are structures which have elements of Bas_1 as their parts. All parts of the structures in Dom_1 are elements of Bas_1. In particular, no element of $Comp_1$ can be a part in a structure. Thus, we can say that the various kinds of structures are all of the same *genus*, for we do not have any structures of structures. Our chemistry example is like this; we have atoms and molecules, and the atoms are the parts of the molecules. Each smallest possible sample is identified with either an atom or with a molecule. We do not (in that example) consider structures whose parts are molecules.

Since $Comp_1$ has structures which are all of the same genus, I will be developing the conditions for one-level microreductions. That is, we jump at most one level, from parts (basic elements) to wholes (compound elements). The conditions I will obtain can be generalized to apply to multi-level microreductions, in which more than one genus of structures is involved. Such reductions can also often be treated as a sequence of several one-level microreductions. It should be clear, after we have treated the one-level case, how one can proceed to these more elaborate situations.

In most theories with structured wholes it will be the case that a given basic element may at one time be uncombined, and at another time combined with other basic elements in a structure. A basic element not combined in a structure is in the *free state*, or is simply called '*free*'. A basic element combined with others in a structure is *bound* or said to be in the *bound state*. In addition, it will usually be the case that the basic elements can exist in various other states under various conditions. Some of these states may be independent of whether or not a basic element is free or bound; others may not. It is assumed that $\mathscr{L}_1$ has appropriate predicates for describing the possible states of the basic elements. Furthermore, since states may depend on certain external environmental conditions, $\mathscr{L}_1$ has appropriate predicates for describing these relevant environmental conditions.

It is now time to consider in more detail some very important features of compound elements. A *compound element* of type C consists of two or more basic elements satisfying a structural description ϕ defined in terms of basic thing-predicates together with appropriate attribute-predicates. A *structural description* ϕ specifies that a finite number of basic elements of certain types are so arranged that a certain structural relationship holds between them. This structural relationship must be definable in terms of basic attribute-predicates. If c is a compound element, the *domain* of c is the set of basic elements composing c. Two compound elements, c_1, c_2, are of the same *kind* (or *type*) iff they satisfy some given structural description. Thus, c_1, c_2 are of the same kind iff (i) there is a one-to-one correspondence between the elements in their domains which correlates basic elements of the same kind, and (ii) the corresponding basic elements in their domains are structurally related in the same way in c_1 and c_2.

Since I do not want to restrict the discussion to any particular variety of structural relation, this characterization of 'same kind' for compound elements has intentionally been stated in very general terms. Although its meaning should be clear, it might be helpful to illustrate it with a simple example. Suppose that we have a *Bas*₁ set consisting of several kinds of balls. Some of the balls have holes, some have pegs, and some have holes and pegs. Let the *type* of a ball be determined by the variety of holes or pegs it has. An obvious, primitive structural relation is the relation, R, such that Rab holds iff ball a has a peg connected to a hole in ball b. In particular, let us suppose that there are four types of balls denoted by the primitive homogeneous thing-predicates, 'P', 'H', 'T', and 'D'. A P-ball has one peg only, an H-ball has one hole only, a T-ball has three pegs symmetrically arranged in a plane with 120 degrees between them, and a D-ball has one hole and one peg on exactly opposite sides of the ball.

Now suppose, for the sake of the example, that a possible structure consists of two or more balls connected in such a way that there are no unused pegs or holes in these balls. Then it is easy to describe these possible structures. For instance, the simplest *kind* of compound element is an L_0-compound, where we have: s is an L_0-compound iff s is a set $\{a, b\}$ such that $Pa \,\&\, Hb \,\&\, Rab$. It is seen that the parts a, b, satisfy the structural description ϕ_0, where ϕ_0 is defined by: $\phi_0\,ab$ iff $Pa \,\&\, Hb \,\&\, Rab$. Similarly, one can define other *linear L_n-compounds*, where $n > 0$. Also one can define compounds which consist of a T-ball with three linear arms extending from it. Thus, in general, the various *kinds* of compound elements are determined in a straightforward way by their domains and the structural relationships holding between the elements in these domains.

This simple ball-and-peg example illustrates a very important condition which it is assumed is satisfied by $\mathscr{L}_1$, namely, that *it is possible in $\mathscr{L}_1$ to define the various possible kinds of compound elements contained in $Comp_1$*. Moreover, these definitions of the various kinds of compound elements are to provide a classification of $Comp_1$ by structures. Therefore, these definitions should partition $Comp_1$ into disjoint subsets such that any two elements in a given subset have the same structure and any two elements in different subsets have different structures. Thus, if ϕ_1, ϕ_2 are two structural descriptions, then either $Ext_{\phi_1} = Ext_{\phi_2}$ or $Ext_{\phi_1} \cap Ext_{\phi_2}$ is empty. But the classification of $Comp_1$ is being accomplished entirely by means of the definitions in terms of structural descriptions. Therefore, the only way we have of identifying or distinguishing kinds of compound elements is by means of structural descriptions. Thus, it must follow solely from the definitions of ϕ_1, ϕ_2 that either $Ext_{\phi_1} = Ext_{\phi_2}$ or $Ext_{\phi_1} \cap Ext_{\phi_2}$ is empty. Hence, if ϕ_1, ϕ_2 are any two structural descriptions, then they must either be analytically equivalent or else analytically disjoint. This implies that two compound elements, c_1, c_2, are of the same kind iff they satisfy analytically equivalent structural descriptions. The latter condition is satisfied iff c_1, c_2 satisfy some given structural description.

Having seen how the predicates for the kinds of compound elements are defined in $\mathscr{L}_1$, it is now necessary to consider how various kinds of attribute-predicates can be defined. In particular, let us begin by considering what a defined classifying attribute-predicate is.

Suppose that ψ is a binary relation defined in $\mathscr{L}_1$, and let E_1 be the natural equivalence relation defined on Bas_1. Then ψ is a *defined classifying attribute-predicate* iff for all basic elements, b_1, b_2, b_1', b_2', if $E_1 b_1 b_1'$ and $E_1 b_2 b_2'$ and $\psi b_1 b_2$ (under specified environmental conditions), then $\psi b_1' b_2'$ (under the specified conditions). In other words, whenever basic elements of certain kinds are related by ψ, under specified conditions, any other basic elements of the same kinds (when taken in the proper order) are also related by ψ under these specified conditions.

This definition is immediately generalizable to n-ary predicates, and to n-ary operations. Furthermore, by this definition any two basic elements are of the same kind iff they have exactly the same classifying attributes, both primitive and defined.

At this point I want to make a very important assumption about the behavior of the basic elements. Suppose that $b_1, \ldots, b_k, \ldots, b_n$ are basic elements of various kinds, and suppose that these basic elements form a compound element of a certain kind under some specified conditions. Also let $b_1', \ldots, b_k', \ldots, b_n'$ be basic elements of the same kinds as the $b_1, \ldots, b_k, \ldots, b_n$, respectively. Then it would be extremely peculiar if the b_i' ($i = 1, \ldots, n$) do not also form the same kind of compound element as the b_i under the same specified conditions. For instance, in chemistry it would be extremely bizarre if some hydrogen and oxygen atoms formed H_2O molecules under certain conditions and other hydrogen and oxygen atoms did not. Therefore, I am going to assume that the basic elements satisfy a *Uniformity Condition*. Let b_i ($i = 1, \ldots, n$) and b_i' ($i = 1, \ldots, n$) be basic elements, and let ϕ be a structural description for a kind of compound element. Suppose further that $E_1 b_i b_i'$ ($i = 1, \ldots, n$). Then, if the b_i satisfy ϕ under specified conditions, then the b_i' also satisfy ϕ under these conditions. Obviously, this Uniformity Condition is simply the requirement that all structural descriptions are defined classifying attribute-predicates. It is a condition which holds in a wide variety of scientific theories. As I stated at the end of Section C, I take it to be a necessary condition for a uniform microreduction.

We must now consider *compound attribute-predicates*, i.e., (defined) attribute-predicates whose extensions are sets of compound elements. Such a predicate can be introduced in the following manner. If c is a compound element, then Pc iff $\psi b_1, \ldots, b_n$, where $b_1, \ldots, b_n$ are elements in the domain of c, and where ψ is a predicate defined in $\mathscr{L}_1$ in terms of basic attribute-predicates possibly together with appropriate basic thing-predicates. For instance, a molecule is *linear* iff the atoms in it are arranged in certain general ways. Of course, in a similar manner, one can define compound n-ary relations and compound operations.

We are primarily interested in compound classifying attribute-predicates. It is easy to see that 'P' is a *compound classifying attribute-predicate* iff ψ is a defined classifying attribute-predicate. In such a case we have substitutivity. If c_1, c_2 are two compound elements of the same kind and if Pc_1, then Pc_2.

We have already seen that two basic elements are of the same kind iff they have exactly the same classifying attributes, both primitive and defined. Furthermore, if b is a basic, and c a compound element, then b and c are clearly of different kinds. Moreover, they differ in at least one classifying attribute, for c satisfies a structural description which b does not, and structural descriptions are defined classifying attributes. Likewise, any two different kinds of compound elements satisfy different, and mutually exclusive, structural descriptions. On the other hand, we saw in the previous paragraph that any two compound elements of the same kind have the same compound classifying attributes. All of this leads to a very important conclusion. From now on, unless otherwise indicated, let us mean by 'classifying attribute' the entire set of classifying attributes, primitive, defined, and (defined) compound. Then we see from this discussion that *any two elements in Dom_1 are of the same kind iff they have exactly the same classifying attributes.* I call this the *General Homogeneity Principle*, and it has just been shown that this principle follows from our characterization of Dom_1. Of course, an important part of this characterization was the Uniformity Condition that all structural descriptions are defined classifying attribute-predicates.

Let us now review what has been said in this section about Dom_1 and $\mathscr{L}_1$. $Dom_1 = Bas_1 \cup Comp_1$, where Bas_1 is nonempty, and $Comp_1$ is nonempty in most microreductions. This section was concerned with microreductions in which $Comp_1$ is nonempty. In such cases it is necessary that $\mathscr{L}_1$ be able to define all of the possible kinds of compound elements in $Comp_1$. These definitions are accomplished by using structural descriptions which describe the structures of the various kinds of compound elements. Moreover, these structural descriptions are classifying attribute-predicates. In addition, other kinds of attribute-predicates will usually be definable in $\mathscr{L}_1$. Some of these will be classifying attribute-predicates, while others may not. However, by the General Homogeneity Principle, we know that any two elements of Dom_1 are of the same kind iff they have exactly the same classifying attribute-predicates. In this way, Bas_1 is partitioned into homogeneous subsets of basic elements, and $Comp_1$ is also partitioned into homogeneous subsets of compound elements. These homogeneous subsets of compound elements are named by various kinds of structural descriptions defined in $\mathscr{L}_1$. Figure 1 is a schematic diagram of Dom_1.

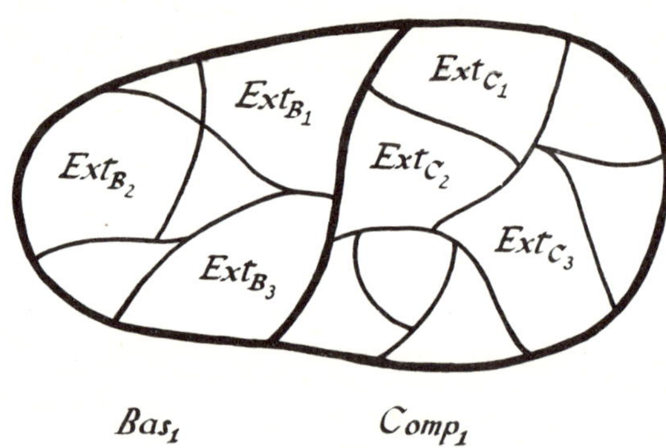

Figure 1. Diagram of Dom_1. Bas_1 is partitioned into homogeneous natural equivalence classes which are named by basic thing-predicates, B_1, B_2, B_3, ... $Comp_1$ is partitioned into homogeneous equivalence classes of structured wholes which are named by compound thing-predicates, C_1, C_2, C_3, ... The C_i are determined by defined structural description predicates, ϕ_1, ϕ_2, ϕ_3, ...

It might appear that all of this discussion about compound elements merely consists of unmotivated technicalities. But this is not the case. In fact, the General Homogeneity Principle is practically taken for granted by many scientists when reductions are performed. Dom_2 is partitioned by E_2 into natural equivalence classes, which are homogeneous. A successful reduction is required, among other things, to provide an understanding of the homogeneity of these classes. The way this understanding is obtained is by identifying these classes with homogeneous classes in Dom_1. In other words, the reduction shows that $Dom_2 \subseteq Dom_1 = Bas_1 \cup Comp_1$. But, of course, homogeneous classes of Dom_2 must be identified with homogeneous classes of Dom_1. Furthermore, a micro-reduction will involve identifying a certain kind of element of Dom_2

with either a certain kind of basic element of Dom_1 or with a certain kind of compound element of Dom_1 which has a characteristic kind of structure. Therefore, in order for the microreduction to account adequately for the homogeneity of a class of Dom_2 elements, it is necessary that the General Homogeneity Principle hold. Our example from chemistry illustrates this point.

Finally, it should be clear why I have required that the various kinds of compound elements and compound attributes be denoted by predicates defined in terms of basic thing- and attribute-predicates. This is necessary because we want the reduction to provide an understanding of a Dom_2 element in terms of its parts (when it is identified with a $Comp_1$ element). The various conditions imposed on $\mathscr{L}_1$ help to accomplish this understanding. If we did not restrict $\mathscr{L}_1$ in this way, then it would be possible for $\mathscr{L}_1$ to contain, for example, a primitive compound attribute-predicate. It could then be charged that such a predicate denotes an "emergent property" not completely reducible to the basic elements and their attributes. A similar point about emergent laws will be discussed in the next section.

E. Laws of T_1

Needless to say, different theories with structured wholes will apply to different domains of elements exhibiting different kinds of behavior. Therefore, it is impossible to describe in detail the forms of the laws of T_1. However, it will perhaps be useful to mention at least a few general types of laws one might expect to find in most T_1's.

First of all, recall that T_1 is formulated in terms of the nonlogical predicates in $\mathscr{L}_1$. Moreover, Bas_1 is partitioned into natural equivalence classes by the natural equivalence relation on Bas_1. Therefore, at the very least we would expect T_1 to contain laws which assert which attributes various basic elements have under specified environmental conditions. Roughly speaking, such laws tell us what *states* basic elements are in under various conditions. In addition, if some basic elements can be transformed into others, then we would expect T_1 to contain laws describing the nature of these transformations and the conditions under which they occur. Also, if basic elements can undergo various transformations or changes of state, then T_1 may also contain laws specifying the *rates* of these changes, i.e., how fast certain changes occur under certain conditions.

In the second place, $Comp_1$ will usually be nonempty. When this is the case, T_1 should also contain laws stating conditions under which various compound elements do or do not exist. These laws are often of very great importance. I have discussed some of their features in Causey (1969b). For our purposes it is sufficient to notice that such laws will usually be required if we are to explain transformations which Dom_2 elements can undergo. In this connection it should also be noted that if elements of Dom_1 can react with one another under certain conditions, then T_1 would normally have to describe these reactions and the conditions under which they occur.

Of course, the compound, as well as the basic elements of Dom_1, exhibit certain attributes under certain conditions. The predicates denoting compound attributes are defined attribute-predicates. But it is not sufficient merely to define these predicates. In addition, T_1 will normally have to contain laws that tell us which attributes the various kinds of compound elements have under specified conditions. T_1 may also contain laws which give the rates of various changes that compound elements can undergo under specified environmental conditions.

Finally, it must be noted that the notion of environmental conditions has intentionally been left rather vague. Environmental conditions might be almost anything. For example, the environment of a chemical reaction could include such things as temperature, pressure, presence of light, presence of mechanical shock, etc. The relevant aspects of the environment of an organism may include all of the above, as well as such things as other organisms, nutrients, pollutants, etc. The relevant aspects of the psychological environment of a human may include his family relations, the social system in which he is found, etc.

Needless to say, $\mathscr{L}_1$ and T_1 will need to make reference to environmental conditions and to state some relations between the elements of Dom_1 and the relevant aspects of the environment. If these aspects include objects, then they would normally be considered elements of Dom_1. Fortunately, environmental conditions can often be stated in general terms by giving the values of various environmental state functions, e.g., temperature, pressure, electric gradient, etc.

The above discussion should have provided a general picture of the type of theory T_1 will usually be. This picture includes laws about both basic and compound elements. However, in order for T_1 to provide a satisfactory reduction of T_2, T_1 must be *fundamentally* a theory about Bas_1 and only *derivatively* a theory about $Comp_1$. This is partly accomplished by the conditions imposed on $\mathscr{L}_1$ and Dom_1. However, these conditions are not sufficient.

For instance, suppose that ϕ is a structural description defined in $\mathscr{L}_1$, and let ψ be a compound attribute-predicate defined in $\mathscr{L}_1$. Also let ϕ define the compound thing-predicate 'C' and let ψ define the compound attribute-predicate 'P'. Now consider the law-sentence which asserts that, for any compound element c,

$$Cc \to Pc. \tag{3.1}$$

The law-sentence (3.1) is analytically equivalent to the law-sentence which asserts that, for any basic elements $b_1, \ldots, b_n$,

$$\phi b_1, \ldots, b_n \to \psi b_1, \ldots, b_n. \tag{3.2}$$

Now, clearly, we do not want (3.1) or (3.2) to be *fundamental* law-sentences of $\mathbf{T}_1$, for we want to understand the behavior of the wholes (compound elements) in terms of the behavior of their parts (basic elements). Thus, a law-sentence such as (3.1) needs to be explained, and thus suitably derived, from some set of fundamental law-sentences about Bas_1. If this is not done, it might be charged that the law is emergent. Therefore, all of the law-sentences of $\mathbf{T}_1$ should be explainable from a set of *fundamental* law-sentences which, in some sense, apply directly only to Bas_1. How can we characterize such a set of fundamental law-sentences?

Unfortunately, there does not seem to be any simple, general answer to this question. But it is possible to mention a few kinds of laws which would be acceptable as fundamental laws. Recall from Section D that a basic element is *free* when it is not combined with other basic elements in a compound element; when it is so combined, we say that it is *bound*. We would then normally expect $\mathbf{T}_1$ to contain some laws of the following type: a free basic element of a certain kind possesses a certain attribute under certain environmental conditions. A law of this kind would seem to be a natural candidate for a fundamental law. Likewise, laws expressing nonstructural relations holding between free basic elements would seem to be fundamental laws.

On the other hand, a law which states that, under certain conditions, certain kinds of basic elements form a kind of compound element with a complex structure, would be a dubious candidate for a fundamental law. For example, in chemistry suppose we used a law which simply states that, under certain conditions, six hydrogen and six carbon atoms form a benzene ring molecule. Such a law would seem extremely *ad hoc* unless it were explainable in terms of some general principles of chemical bonding. It is largely for this reason that chemists sought long and hard for a general theory of the chemical bond in terms of which they could explain and predict the various possible kinds of molecules which form under different conditions. Thus, we do not want *ad hoc* laws about the existence of particular kinds of compound elements. However, it is clear that T_1 will have to have some laws about the kinds of structural relationships which are possible. We recall from the previous section that some structure-predicates will be in $\mathscr{L}_1$ or be definable in $\mathscr{L}_1$. By the use of these predicates T_1 could contain a few very general, fundamental laws about the kinds of general structural relationships which can hold under different conditions. Then from these laws, plus more specialized laws about the attributes of various kinds of basic elements, one could *derive* specialized laws about the existence of various kinds of compound elements. This is the kind of procedure which is used in modern molecular structure theory. It is probably the best one can hope to do.

After one has accounted for the existence of the different kinds of compound elements, then he can use their structural descriptions plus other information to explain law-sentences like (3.1). Hence, such law-sentences will be derivative rather than fundamental. In fact, their derivation is very important in the reduction of T_2. I will discuss such derivations in detail in Chapter 4. For the moment, it should be fairly clear that all derivative law-sentences of T_1 should be explainable from a subset of fundamental law-sentences of T_1 which apply directly only to Bas_1. The general character of these fundamental law-sentences should also be fairly clear. In the next chapter I will begin to describe some of the details of microreductions.

MICROREDUCTIONS: SET THEORETICAL FORM

It has long been recognized that in a nontrivial reduction of T_2 to T_1, T_1 must be supplemented with a set of connecting sentences in order to make possible the explanation of the law-sentences of T_2. In the case of microreductions, it has been recognized that some of these connecting sentences should be some kind of identities between the elements in Dom_2 and elements in Dom_1. However, most investigators have also thought that at least some of the connecting sentences are certain types of law-sentences, and for this reason these connecting sentences are often called "bridge laws". I will eventually argue that all connecting sentences must be identities; some will be thing-identities and some will be attribute-identities. However, in this chapter I will develop, as far as is possible, conditions for microreductions in set theoretical terms.

Section A will examine the connections needed between the thing-predicates of the two theories. Section B will be concerned with some problems involved in the explanations of the law-sentences of T_2. Section C will examine the possibility of connecting sentences which are nomological co-extensionalities between the attribute-predicates of the two theories.

The work of this chapter will produce, I believe, the best possible conditions for uniform microreductions which are formulated in a set theoretical framework. In Chapter 5 I will then argue that these conditions are not completely adequate. They will then be replaced by a set of conditions in which all connecting sentences are identities.

A. THING-IDENTITIES

In the classical model of reduction presented in Ernest Nagel (1961, pp. 336–366), the aim is to explain T_2 by T_1. As Nagel showed, except for trivial cases, it is necessary to supplement T_1 with a set, **B**, of connecting sentences in order to accomplish the reduction. This is necessary in order to make possible the derivation of the law-sentences of T_2 because $\mathscr{L}_2$ has terms not in $\mathscr{L}_1$. Nagel allowed (p. 355fn.) that some connecting sentences might be biconditionals, and some might be one-way conditionals. At one time Hempel (1966, p. 105) did likewise.

On the other hand, some other investigators, Woodger (1952, pp. 271–272), Adams (1959, p. 256), Schaffner (1967, p. 144) have presented conditions for reduction which require the connecting sentences to be some kind of identities or biconditionals. Unfortunately, these investigators have relied largely on set theoretical languages, and have therefore not distinguished genuine thing- and attribute-identities from nomological co-extensionalities. They were largely concerned with very general logical conditions for all sorts of reductions. As we will see, identities naturally receive more attention when microreductions are examined in some detail. Fortunately, since most reductions seem to be microreductions, indeed, uniform microreductions, the conditions I obtain should be of considerable, general interest.

To begin, consider the fact that a microreduction somehow essentially involves the identification of the elements of Dom_2 with elements in Dom_1. Thus, at the very least, we should expect some elements of **B** to be some kind of identities. Let us see how they might be expressed in a set theoretical language.

Let 'W' be a homogeneous thing-predicate of $\mathscr{L}_2$, i.e., 'W' is a thing-predicate which names a homogeneous subset of Dom_2. Then all of the elements of Ext_W are the same except possibly for differences in their nonclassifying attributes. Thus we would want to identify an arbitrary element of Ext_W with a certain kind of basic or compound element of Dom_1. Hence, we should have a connecting sentence of the form

$$Wx \leftrightarrow (\exists y)(\alpha y \;\&\; y = x), \tag{4.1}$$

where α is some homogeneous thing-predicate in, or defined in, $\mathscr{L}_1$, i.e., α is a predicate naming a kind of basic or compound element of Dom_1. In writing (4.1) I follow the common convention that free variables are understood to be universally quantified. Thus, (4.1) tells us that anything, x, which is a W, is identical to some y which is an α. But (4.1) is unnecessarily complicated, for it is set theoretically equivalent to

$$Wx \leftrightarrow \alpha x. \tag{4.2}$$

Now (4.2) tells us that $Ext_W = Ext_\alpha$. Any connecting sentence like (4.2), which is a co-extensionality between a homogeneous thing-predicate of $\mathscr{L}_2$ and a homogeneous thing-predicate in, or defined in, $\mathscr{L}_1$, is a *Thing*

Identity Connecting Sentence (a TICS). Connecting sentences of this kind underlie (or may someday underlie) identity claims such as:

a. an argon atom is (identical with) a certain kind of structure composed of electrons, protons, and neutrons,

b. a sample of argon gas is a collection of randomly distributed argon atoms,

c. a (macroscopic) ice crystal is a certain kind of crystalline lattice of H_2O molecules,

d. a certain kind of gene is a certain portion of a DNA molecule,

e. an amoeba is a certain structure composed of the parts ...,

f. a coral reef is a structure composed of coral animal skeletons,

g. a red light wave is an electromagnetic wave of such and such frequency,

h. a lightning bolt is an electrical discharge,

i. a human being is a ...,

j. a clique is a certain kind of social structure.

Now it should be obvious that TICS's are essential to microreductions. In fact, since we want $Dom_2 \subseteq Dom_1$, we want each homogeneous thing-predicate of $\mathscr{L}_2$ to be *reduced* to a homogeneous thing-predicate of $\mathscr{L}_1$. This can be accomplished if **B**, the set of connecting sentences of the reduction, has a TICS for each such predicate of $\mathscr{L}_2$. Moreover, **B** needs at most one TICS for each of these predicates. For suppose that **B** has, in addition to (4.2), the TICS

$$Wx \leftrightarrow \alpha'x. \tag{4.3}$$

Then one can derive

$$\alpha x \leftrightarrow \alpha'x, \tag{4.4}$$

which is a true sentence of any adequate T_1. To see that this is the case, we consider the following.

Clearly, both α, α' denote basic kinds of elements, or both denote kinds of compound elements. If α, α' are both basic thing-predicates of $\mathscr{L}_1$, then (4.4) is either analytic (a definition) or a synthetic identity. If α, α' denote kinds of compound elements, then (4.4) is an analytic equivalence of two structural descriptions (see Chapter 3, Section D).

We see that in all cases, (4.4) is not only extensionally true, but it is also either analytic or a synthetic thing-identity. If it is analytic, then it is clearly in T_1 because it is a mere consequence of the language $\mathscr{L}_1$. On the other hand, suppose that (4.4) is a synthetic identity. Then α, α' merely denote the same homogeneous subset of Dom_1. Thus, anything which is an α would exhibit exactly the same attributes and lawful behavior as anything which is an α'. This fact, and hence also the identity of α and α', should be manifestly true in any adequately formulated T_1. Therefore, in all cases (4.4) is in T_1, so (4.3) is not needed in **B**. We see in addition, that if a sentence of the form (4.3) is true, then its truth is consistent with that of (4.2).

This discussion shows that **B** needs at most one TICS for each primitive homogeneous thing-predicate of $\mathscr{T}_2$ which is not in $\mathscr{T}_1$. In a non-trivial microreduction it is unlikely that any TICS will be analytic, and if any are, they will be trivial and will be ignored here. The remaining TICS's may be either nomological co-extensionalities or synthetic thing-identities. I use the term 'TICS' because I will argue in Chapter 5 that they are synthetic thing-identities. For the present, I will continue to interpret them extensionally.

B. Explanations of the Law-Sentences of T_2

In this section I will discuss very briefly how the law-sentences of T_2 will be explained in the course of the reduction of T_2 to T_1. I will only consider a very simple type of T_2-law-sentence, and I will only outline a possible derivation of such a law. However, the discussion will illustrate several important points about the use of connecting sentences and about the role of structural descriptions in microreductive explanations.

Let 'W' be a homogeneous thing-predicate of $\mathscr{L}_2$, let 'A' be a one-place classifying attribute-predicate of $\mathscr{L}_2$ which is not in $\mathscr{L}_1$, and let 'E' be a predicate describing certain environmental conditions. We can, without affecting our present purpose, assume that 'E' is in both $\mathscr{L}_1$ and $\mathscr{L}_2$. Let us now assume that T_2 contains the fundamental law-sentence

$$(Wx \ \& \ Ex) \rightarrow Ax. \tag{4.5}$$

Obviously in a reduction of T_2 one must explain, and he needs only to explain, the fundamental law-sentences of T_2. Therefore, we need to consider how a law-sentence such as (4.5) can be explained. We must also keep in mind that the explanatory derivation must be in the spirit of the overall strategy of the microreduction. Therefore, it is natural to identify W's with some homogeneous class of Dom_1, and then to try to explain why W's exhibit A under condition E in terms of this homogeneous class of Dom_1 and its characteristics.

If the strategy just described is followed, one is likely to come up with a derivation of this kind:

$$Wx \leftrightarrow Cx, \tag{4.6}$$

$$(Cx \ \& \ Ex) \rightarrow Px, \tag{4.7}$$

$$Px \rightarrow Ax, \tag{4.8}$$

therefore,

$$(Wx \ \& \ Ex) \rightarrow Ax. \tag{4.5}$$

For the sake of an interesting example, let us suppose that 'C' is a compound thing-predicate defined in $\mathcal{L}_1$, and that 'P' is a compound attribute-predicate defined in $\mathcal{L}_1$.

It is clear that the conclusion of this argument follows from the premises. But what are the premises? First of all, we see that (4.6) is a TICS which at least extensionally identifies W's with C's. Also it is clear that (4.8) is some kind of connecting sentence, or at least a logical consequence of a connecting sentence. Finally, (4.7) is clearly a derived law-sentence of $\mathbf{T}_1$. Therefore, the argument seems quite simple, consisting of two connecting sentences and a derived law-sentence of $\mathbf{T}_1$. Later, in Section C, I will have much more to say about connecting sentences similar to (4.8). For the moment, we need to consider (4.7) more carefully.

It should be quite clear that, unless (4.7) can be adequately explained in $\mathbf{T}_1$, it will be mysterious why C's have P under E. Moreover, if this mystery is not removed, it will carry over into the reductive explanation of (4.5). Indeed, if (4.7) is not derived from fundamental laws of $\mathbf{T}_1$, it could be charged that the property named by 'P' is an emergent property of C's. Furthermore, the aim of a microreduction is to provide an understanding of why W's have A, in terms of the decomposition of W into its parts. Clearly, in order to obtain such understanding, we need more than merely the argument given above. In fact, what we need to do is to interpolate into the above argument a derivation of (4.7). All we can do here is give a rough outline of the general form of such a derivation. I have given such an outline previously in Causey (1969b), but it will be presented here in a slightly different manner.

First of all, we need the structural description, ϕ, of the C's. Thus, we have Cx iff x is a set of basic elements, $b_1, \ldots, b_n$ which satisfy ϕ. Also we need the definition, ψ, of the compound predicate 'P'. Thus, Px iff $\psi b_1, \ldots, b_n$. Now consider the following:

$$Cx \ \& \ Ex. \tag{4.9}$$

$$Cx \leftrightarrow \phi b_1, \ldots, b_n, \quad \text{where} \quad \{b_1, \ldots, b_n\}$$
$$\text{is the domain of } x. \tag{4.10}$$

$$Px \leftrightarrow \psi b_1, \ldots, b_n, \quad \text{where} \quad \{b_1, \ldots, b_n\}$$
$$\text{is the domain of } x. \tag{4.11}$$

Now note that (4.9) merely says that the compound element x is under condition E. Then, by the *definition* (4.10), we see that $b_1, \ldots, b_n$ are in the combined environment of E plus that determined by their being in the structure ϕ. Thus, in the derivation of (4.7) we use this information plus fundamental law-sentences of $\mathbf{T}_1$ which describe the behavior and properties of the basic elements $b_1, \ldots, b_n$ in order to derive that $\psi b_1, \ldots, b_n$. Having this, we use (4.11) to conclude Px. Therefore, all we assume is (4.9) plus fundamental law-sentences of $\mathbf{T}_1$, since (4.10) and (4.11) are merely definitions. Therefore,

$$(Cx \ \& \ Ex) \rightarrow Px \tag{4.7}$$

is explainable by the fundamental law-sentences of $\mathbf{T}_1$, and therefore it is a derivative law-sentence of $\mathbf{T}_1$.

This short discussion should give a pretty good idea of the importance of structural descriptions in explaining the attributes of wholes. When one thinks of chemistry, this should be no surprise. The structures of molecules are constantly involved in explanations. For instance, the fact that benzene is a ring molecule is used in explaining many of benzene's properties. Indeed, this ring structure was initially hypothesized in order to account for many of its properties. We see that the structural descriptions of compound elements play a crucial role in microreductive explanations. However, I still have not said all that needs to be said about (4.8). In the next section I will argue that connecting sentences involving attribute-predicates should *at least* be correlations. Finally, in Chapter 5, I will argue that *all* connecting sentences should be identities.

C. ATTRIBUTE-CORRELATIONS

We have seen that TICS's (which are at least correlations) are necessary because at least part of what is meant by a microreduction is that it is a reduction which asserts that Dom_2 is a subset of Dom_1. However, it is not obvious what kind of connecting sentences are needed for the attribute-predicates of $\mathscr{L}_2$. In this section I will argue that they also must *at least* be co-extensionalities. Of course, nontrivial co-extensionalities will not be analytic, and we would not want connecting sentences which are accidentally true. Therefore, within a set theoretical interpretation of theories, we would expect biconditional connecting sentences to be *at least* nomological co-extensionalities. Nomological co-extensionalities are often called 'correlations'. This section is therefore concerned with reasons why **B** should *at least* have attribute correlation connecting sentences.

In recent years there has been increasing interest in the logical form of connecting sentences. For instance, Sklar and Hempel have (apparently independently) given one reason for requiring a biconditional connecting sentence for each predicate of $\mathscr{L}_2$ which is not in $\mathscr{L}_1$. Their reasoning goes roughly as follows. Suppose that we are able to derive T_2 from $T_1 \cup B$, where **B** is a set of connecting sentences. Suppose further that not all predicates of $\mathscr{L}_2$ are extensionally equivalent (in $T_1 \cup B$) to predicates in, or defined in, $\mathscr{L}_1$. Then, even under a set theoretical interpretation of theories, all that we have accomplished is the reduction of T_2 to $T_1 \cup B$, and $T_1 \cup B$ is an $\mathscr{L}_2$-theory (say, a biological theory) rather than an $\mathscr{L}_1$-theory (say, a chemical theory) because $T_1 \cup B$ still makes essential use of some $\mathscr{L}_2$-predicates. Paraphrasing Hempel (1969, p. 189), $T_1 \cup B$ is an $\mathscr{L}_2$-theory because it has $\mathscr{L}_2$-law-sentences, i.e., law-sentences involving $\mathscr{L}_2$-predicates which are not extensionally equivalent (in $T_1 \cup B$) to predicates definable in $\mathscr{L}_1$. Of course, if for each $\mathscr{L}_2$-predicate there is in $T_1 \cup B$ a biconditional *reducing* this predicate, then this biconditional would function as an "extensional definition" of the predicate. Thus, under the set theoretical interpretation of theories, the use of this predicate would be avoidable. It should be observed that this line of reasoning does not require that the biconditionals be in **B**; it only suggests that they should be in $T_1 \cup B$.

Sklar (1967, pp. 119–120) presents another argument similar to Hempel's, but Sklar seems to interpret his argument as indicating that reductions require identifications of elements of Dom_2 with elements of Dom_1. Moreover, he allows that there might be some connecting sentences which are one-way conditionals (although his examples of these seem to be in error, for when properly formulated I believe that they turn out to be biconditionals).

I will call the argument presented above the 'Sklar-Hempel argument', since it is at least suggested by their writings, and since they seem to be the first investigators who have tried to give an argument to justify the need for biconditional connecting sentences. Needless to say, it is not a powerful argument, but it does show that biconditionals would be nice to have in order to achieve a genuine reduction to an $\mathscr{L}_1$-theory rather than merely to an $(\mathscr{L}_1 + \mathscr{L}_2)$-theory. Perhaps we can find a more compelling reason for requiring them.

Because of the work of Section A we now only need to concern ourselves with $\mathscr{A}_2$, in particular, with the primitive attribute-predicates of $\mathscr{L}_2$. Therefore, let 'A' be a primitive predicate in $\mathscr{A}_2$ which is not in $\mathscr{A}_1$. To avoid unnecessary complications, assume that 'A' is a one-place predicate standing for a *property* of certain elements of Dom_2. The argument to be presented is generalizable to attributes which are relations or quantities.

In the case of A, there are some environmental conditions under which some elements of Dom_2 have A. Therefore, let a_1 be some kind of element of Dom_2 which has A under some environmental conditions, say E. From Section A we have that any set $\mathbf{B}$ of connecting sentences will have TICS's applying to the elements of Dom_2. Therefore, $a_1 = e_1$, where e_1 is some kind of basic or compound element of Dom_1. Therefore, any suitable $\mathbf{T}_1 \cup \mathbf{B}$ will assert that, under E, Ae_1, if e_1 is the appropriate kind of element.

Now any adequate microreduction should be able to give some kind of account of why it is that Ae_1. In effect, we saw in the previous section how such an account would be given when A is a classifying attribute. But even for an arbitrary attribute A we would at least want a condition defined in $\mathscr{L}_1$ which is sufficient for e_1 to have A. For if we did not have such a *sufficient* condition, then it could be claimed that A is an emergent property of e_1, and thus it could also be claimed that $\mathbf{T}_1$ is not an empirically complete theory about Dom_1. It may be that a mere sufficient condition is still not entirely adequate for explaining why Ae_1. But, at any rate, it seems to me that having a sufficient condition defined in $\mathscr{L}_1$ is at least necessary for avoiding the charge that A is emergent. Thus, at the very least, we will need in $\mathbf{T}_1 \cup \mathbf{B}$ a law-sentence which states that

$$\delta_1 x \rightarrow Ax, \tag{4.12}$$

where δ_1 is defined in $\mathscr{L}_1$, and $\delta_1 e_1$ under E.

Now suppose that there are other kinds of elements, $a_2, a_3, \ldots,$

$a_i, \ldots$, of Dom_2 which also have A under various environmental conditions. Then, for each a_i, there is an e_i of Dom_1 such that $a_i = e_i$. Therefore, for each i, Ae_i. Thus, as with e_1, we will want sufficient conditions covering all of the e_i. Thus, we will have a list of conditions, $\delta_1, \ldots, \delta_i, \ldots$, defined in $\mathscr{L}_1$ which are sufficient for the e_i having A. In other words, for each i, we have that

$$\delta_i x \to Ax, \tag{4.13}$$

where, for each i, $\delta_i e_i$ under the environmental conditions under which Aa_i, respectively.

If the list of δ_i's is finite, then the disjunction of these δ_i is a sufficient condition for A which covers all of the e_i. On the other hand, if the list of δ_i's is infinite, then in order to have a practical way of stating a sufficient condition for an arbitrary e_i's having A, we will need to have defined in $\mathscr{L}_1$ a condition δ such that

$$Ext_\delta = \bigcup_{i=1}^{\infty} Ext_{\delta_i}. \tag{4.14}$$

Furthermore, since $\mathscr{L}_1$ contains set theoretical terminology, if need be we can always use (4.14) as a definition in $\mathscr{L}_1$ of δ. Having δ we can now write

$$\delta x \to Ax, \tag{4.15}$$

where, for each i, δe_i under the appropriate environmental conditions. It should also be noted that (4.15) is like (4.8).

Now, if we know, or have reason to believe, or are willing to hypothesize that there are no elements which have A other than those which satisfy δ, then we would write (4.15) as the biconditional,

$$\delta x \leftrightarrow Ax. \tag{4.16}$$

On the other hand, if we do not feel confident of this biconditional, and instead we feel that there are, or might be, residual elements having A and not satisfying δ, then we would not feel confident that we have a sufficient condition for an *arbitrary* element's having A. This would certainly cause us to doubt whether we have indeed found an adequate account of *why* an arbitrary element has A if it in fact does have A. Furthermore, this would also leave open the possibility that there are other elements which have A such that for them A is an emergent property. Thus, on closer inspection, I think that it becomes clear that we would not be satisfied with a mere sufficient condition for A. Instead we would want to have an entire biconditional such as (4.16), and have justification that it is not an accidental co-extensionality.

This argument has shortcomings, but together with the Sklar-Hempel argument, it gives fairly compelling reasons why $T_1 \cup B$ should *at least* have nomological co-extensionalities for the predicates of $\mathscr{A}_2$. These co-extensionalities are attribute correlation law-sentences.

Of course, we are still not done, for we have only shown that the correlations should be in $T_1 \cup B$. But now it is easy to see that we might as well require that all of these correlations be in B. For if they are not all in B, but are in $T_1 \cup B$, we can always form a new B' such that all of the correlations are in B'. To get B' we need only add to B any of the appropriate biconditionals not already in B. Thus, we can assume that all of the correlation biconditionals reducing the attribute-predicates of $\mathscr{L}_2$ are in B.

Now B has TICS's reducing the thing-predicates of $\mathscr{L}_2$ and correlations reducing the attribute-predicates of $\mathscr{L}_2$. Suppose B has some other kind of law-sentence λ which is neither a TICS nor one of these correlations. Then we can form λ^* by replacing all occurrences of $\mathscr{L}_2$-predicates in λ by their $\mathscr{L}_1$-equivalents. Then λ^* will be in any *adequate* T_1, for it is a law-sentence expressed in $\mathscr{L}_1$ and derivable in $T_1 \cup B$. Surely, any such law-sentence should be a law of T_1, and so if it is not initially in T_1, it can be added to it to form a slightly strengthened T_1. Thus, we might as well assume that λ^* is in T_1. But then B does not need to contain λ, for λ can be derived from λ^* plus the appropriate correlations and TICS's which are in B. Thus, B need only contain correlations and TICS's. This kind of argument involving λ, λ^* has been used previously by Kemeny and Oppenheim (1956).

We saw in Section A that B needs at most one TICS for each homogeneous thing-predicate of $\mathscr{L}_2$ which is not in $\mathscr{L}_1$. Similarly, B needs at most one correlation reducing each primitive predicate of $\mathscr{A}_2$ which is not in $\mathscr{A}_1$. Henceforth, I will call a correlation between an attribute-predicate of $\mathscr{L}_2$ and an attribute-predicate in, or defined in, $\mathscr{L}_1$ an '*Attribute Correlation Connecting Sentence*' (ACCS). This is because it correlates an attribute named by a predicate of $\mathscr{L}_2$ with one named or defined in $\mathscr{L}_1$.

Up to this point I have argued that B should consist of two kinds of connecting sentences, which are TICS's and at least ACCS's. The qualification "at least" is a result of the fact that I have assumed that the predicates in $\mathscr{L}_1$ and $\mathscr{L}_2$ are given a set theoretical interpretation.

The arguments in this chapter show that when T_1 and T_2 are formulated under this interpretation, then **B** should consist of TICS's and ACCS's. This is the best one can hope to do using set theoretical languages. The reason for this lies in the fact that set theoretical languages do not allow us to refer to attributes or to state attribute-identities. To some extent the same is true for kinds and thing-identities. Thus, the arguments in this chapter are limited by the assumption that the theories use set theoretical languages. In the next chapter I will show that this is a very serious limitation. As a result the reduction conditions developed in this chapter turn out to be inadequate, although they are partially correct and they are the best one can develop using a set theoretical interpretation of the predicates in the theories.

In Chapter 5 it is argued that the TICS's are not arbitrary co-extensionalities between thing-predicates, but are rather thing-identities, i.e., genuine identities of kinds. It is also argued in the next chapter that ACCS's are not adequate to accomplish a reduction. Instead an adequate reduction requires attribute-identities reducing the attribute-predicates of $\mathscr{A}_2$. From Chapter 2 we know that no ACCS is an attribute-identity. Thus, when I now say that **B** should consist of TICS's and at least ACCS's, I mean that these conditions are the best one can hope for using set theoretical languages. But I also mean to imply that an adequate reduction really requires the stronger condition that **B** consist of only thing-identities and attribute-identities, and that this stronger condition cannot be stated within set theoretical languages. Consequently, the arguments presented in the next chapter not only lead to stronger and more adequate reduction conditions; these arguments also provide additional reasons why set theoretical languages are inadequate for at least some scientific purposes.

CHAPTER 5

MICROREDUCTIONS WITH IDENTITIES

This chapter will conclude my detailed analysis of the conditions for uniform microreductions. In Section A I will discuss thing-identity connecting sentences. In Section B I will argue that the ACCS's of Chapter 4 should be replaced by attribute-identities. Section C summarizes the conditions for uniform microreductions and includes a discussion of some further consequences of these conditions. In Section D I will consider some possible objections and problems concerned with the use of identities as connecting sentences. Finally, in Section E I will apply my conditions for microreductions to some questions about modifications of the theories and about trivial microreductions.

My views concerning the use of attribute-identities as connecting sentences were originally presented in Causey (1972a). The present chapter contains a more highly developed and much more extensive discussion of these views. This chapter uses some technical terminology which is different from that used in Causey (1972a).

A. THING-IDENTITIES

A nontrivial TICS is not analytic. Therefore, under a set theoretical interpretation, a TICS is usually a synthetic co-extensionality of the form

$$Wx \leftrightarrow \alpha x, \tag{5.1}$$

where W is a homogeneous thing-predicate of $\mathscr{L}_2$, and α is a homogeneous thing-predicate in, or defined in, $\mathscr{L}_1$. Of course, α may denote either a homogeneous subset of Bas_1 or of $Comp_1$. Suppose that (5.1) is not analytic. Moreover, in a satisfactory microreduction it should not be accidentally true. Then, if the predicates of the theories are interpreted set theoretically, then it would appear that (5.1) would be a nomological co-extensionality. Yet, it can be shown that (5.1) is not a law-sentence.

If (5.1) were a law-sentence, then W would be nomologically correlated with α. Yet, it does not even seem to make sense to say that two homogeneous thing-predicates are merely correlated together. For example, it does not seem to make sense to assert that an empirically smallest sample of water is correlated with an H_2O molecule. Furthermore, if (5.1) were a law-sentence, then the correlation which it would assert would be a causal sentence, i.e., subject to a causal explanation. But suppose someone were to ask: "Why is an H_2O molecule correlated with an empirically smallest sample of water?" We would probably say, "An H_2O molecule is *not* correlated with an empirically smallest sample of water, rather an H_2O molecule *is* an empirically smallest sample of water." This answer is obviously not a causal explanation, but rather an assertion of identity. I have argued in Chapter 2 that identities are not subject to causal explanations, and hence are noncausal sentences.

But why should (5.1) be a noncausal sentence? The reason for this lies in the fact that the thing-predicates involved function simply as *names* for homogeneous natural equivalence classes (kinds) of elements. As names, they do not refer to any attributes these elements might have as a matter of empirical fact. This is even true of compound thing-predicates defined with the help of structural descriptions. The structural description of a compound element is used as part of the definition of this element. Therefore, it is analytically true that a certain type of compound element has the structure it has, and hence we cannot causally explain why it has that structure.

Now, relative to a given set of classifying attributes, the various *kinds* of elements are the various homogeneous natural equivalence classes, and the latter are denoted by the thing-predicates of the theory. Therefore, if two such thing-predicates, even from different theories, are co-extensional, as in (5.1), there is no way such a co-extensionality could be causally explained. This is because the co-extensionality merely asserts that the two thing-predicates name the same thing, i.e., a certain homogeneous natural equivalence class. Therefore, (5.1) should not be interpreted as a causal law-sentence, but rather as a sort of identity, namely, a thing-identity, which asserts the identity of two kinds of things.

One might agree at this point that (5.1) is noncausal, but also remark that this entire discussion has relied on a set theoretical analysis of kinds. It could then be asserted that there is no need to introduce an ontology including *kinds*, and that, at least so far, set theoretical languages have not been shown inadequate for microreductions. Although there is some plausibility in this remark, it overlooks a very important fact.

Homogeneous thing-predicates are not arbitrary names for arbitrary sets; they are very special names for very special kinds of sets. This fact is manifest from the detailed analysis, presented in Chapter 3, of homogeneous thing-predicates. Consideration of their special nature will help one to see that (5.1) is noncausal and is an identity. The following examples will make this more apparent.

In Putnam (1969) it is claimed that the relation of reduction is not extensional. To show this Putnam claims that 'water' is co-extensional with '(H_2O ∨ Unicorn)', although 'water' only reduces to 'H_2O'. Putnam thinks that the first co-extensionality is probably accidental, and that the second one, used in the reduction, is nomological. The status of the first co-extensionality is doubtful; but according to my analysis, the second is a noncausal thing-identity when properly formulated. Unfortunately, Putnam's example is unnecessarily complicated because it mixes unicorns into chemical theory. Therefore, let us consider a similar, but better, example.[1]

We saw in Chapter 3 that T_1 should be able to explain which kinds of structures can or cannot exist under various conditions. It is possible that some kinds of structures may, as a matter of empirical fact, not exist under any conditions. Let us suppose, for the sake of an example, that the figure-8-shaped molecule, Ne_{13}, does not exist under any conditions (as far as I know this is true). Also suppose that the atomic-molecular theory (the T_1 of this example) can explain adequately why this figure-8-shaped molecule does not exist. Let 'S' be a structural description of this molecule. Then it is a derivative law in T_1 that 'S' is co-extensional with the empty set. Then it is also a derivative law in T_1 that '(S ∨ H_2O molecule)' is co-extensional with the predicate 'H_2O molecule'. Finally, let P be some unique property of H_2O molecules, so that we have the law-sentence

If x is an H_2O molecule, then Px.   (5.2)

Then we can substitute into (5.2) the term '(S ∨ H_2O molecule)' for the term 'H_2O molecule' to obtain a new law-sentence. However, this new law-sentence is clearly not n-equivalent to (5.2). To explain why (5.2) is true, we need only explain why H_2O molecules have P. To explain this new law-sentence we must *also* explain why S is empty, i.e., why S's do not exist.

This example shows that we can have a nomological co-extensionality between a homogeneous thing-predicate and a logical combination of homogeneous thing-predicates. Substitution into (5.2) does not leave

invariant the law expressed because we are using a nomological co-extensionality and not a noncausal identity.

This example shows that merely knowing that two predicates are co-extensional does not in general enable us to know whether or not we have a causal or a noncausal co-extensionality. Thus, we see that the fact that (5.1) is noncausal is the result of its logical form plus the special nature of homogeneous thing-predicates. There is thus some justification in saying that (5.1) is a thing-identity while also insisting that

$$(Sx \lor x \text{ is an } H_2O \text{ molecule}) \leftrightarrow x \text{ is an } H_2O \text{ molecule} \quad (5.3)$$

is not a thing-identity, although it expresses an identity of extensions. (5.3) is a causal sentence. Its causal explanation is obviously related to the causal explanation of why there are no S's.

It might still be claimed that the substitution into (5.2) resulted in a different law because 'H_2O molecule' has a meaning different from that of '$(S \lor H_2O$ molecule)'. Unfortunately, it is difficult to rule out meaning as a factor in this particular example. However, I do not believe that meaning is the most relevant factor. Recall that the conditions NE and EE, introduced in Chapter 2, do not require sameness of meaning or analytic equivalence. Furthermore, (5.1) is a synthetic identity and the predicates in it are not analytically equivalent.

In conclusion, I believe that this section has at least shown the following: (i) TICS's are noncausal because of their logical form plus the special nature of homogeneous thing-predicates; (ii) TICS's can be distinguished from causal, nomological co-extensionalities such as (5.3); and therefore (iii) we are justified in interpreting TICS's as thing-identities, expressing identities of kinds, even though kinds might be subject to an extensional analysis.

For the purposes of reduction, the most important property of TICS's is the fact that they are noncausal. This will become clear in the next section. I will also show in that section that attribute-identities are needed for reductions, and that they are very different from attribute correlation law-sentences.

B. ATTRIBUTE-IDENTITIES

Recall the attributes, *sink* and $den(x) > den(H_2O)$. In Chapter 2 I showed that these attributes are nomologically co-extensive, but not identical. The connection between these two attributes is an attribute-correlation law. In Chapter 4 I argued that the connecting sentences of a micro-reduction, when interpreted set theoretically, should *at least* be TICS's and ACCS's. Unfortunately, if these connecting sentences are law-sentences, then the microreduction is seriously inadequate because T_2 will not have been explained solely in terms of T_1. We will have explanatory derivations of the fundamental law-sentences of T_2 which use explanans sentences from $T_1 \cup B$, where B is a set of law-sentences. The aim of a microreduction is to obtain understanding of T_2 in terms of the laws of T_1, but we will, at best, have obtained understanding of T_2 in terms of the laws of $T_1 \cup B$. This is highly unsatisfactory, and it suggests that we really need a set of noncausal connecting sentences. Noncausal sentences are not laws, and they are not subject to causal explanations. If B were a set of noncausal sentences, then the only laws which the microreduction would use in the reductive explanations of the T_2-laws, would be the laws of T_1. This is precisely the goal we wish to obtain. Therefore, we should consider the possibility of obtaining a set of non-causal connecting sentences.

In Section A of this Chapter I have already argued that the TICS's are noncausal, and that they can be interpreted as thing-identities. Therefore, the TICS's in B cause no problem for they are automatically noncausal. Furthermore, it is quite natural that they should be thing-identities since a microreduction is intended to involve the identification of the kinds of things in Dom_2 with kinds of things in Dom_1. In fact, it has long been accepted that microreductions identify *things*. It is therefore natural to suggest that microreductions should also identify attributes. Yet, it has long been doubted that this is possible. I believe that this doubt has largely resulted from the traditional view that attribute-identities must be analytically true. Since it was recognized that nontrivial connecting sentences are synthetic, it was thought that the supposedly analytic attribute-identities could not be nontrivial connecting sentences. This is a rather abstract reason for doubting the possibility of attribute-identity connecting sentences. This abstract doubt might have disappeared sooner had the shortcomings of ACCS's been noticed. Let us therefore consider an example.

It is well known that light can be plane-polarized by passing it through a suitable optical filter (such as the Polaroid lenses in some sunglasses). This kind of polarization has the result that the light waves vibrate in a certain plane, rather than in all directions through a point, as is the case with unpolarized light. Now some substances have the property of rotating this plane either to the right or to the left. Such substances are called 'optically active'. For instance, Pasteur, in a celebrated experiment, showed that there are two optically active kinds of tartaric acid, one which rotates the plane to the left, and one which rotates the plane to the right; see Partington (1953, Vol. 4, p. 295). In fact, any optically active compound exists in at least two optically active forms, one right-rotating and the other left-rotating. It is also known that a molecule which rotates polarized light one way, or the other, has a molecular structure which is dissymmetric, i.e., the molecule is not superimposable, by motions in three dimensions, on its mirror image. Thus, the two optically active kinds of tartaric acid have molecular structures which are non-super-imposable mirror images of each other. (Thus, strictly speaking, they are different compounds.) Nowadays, it is accepted as a general correlation that a molecule is optically active if and only if it has a dissymmetric molecular structure. Thus, if x ranges over molecules, and hence also over ESS of compounds, then for every x,

$$Op.Act.(x) \leftrightarrow Dissym.(x). \tag{5.4}$$

However, I do not believe that anyone would feel intuitively that the property of optical activity of an empirically smallest sample is identical with the property of having a dissymmetric molecular structure. Indeed, I doubt that anyone would identify a dispositional property with a topological or geometrical feature of a structure. In fact, (5.4) is an attribute-correlation law-sentence; it expresses a nomological co-extensionality. Thus, on the basis of Chapter 4, it would be an acceptable ACCS provided that '$Op. Act.$' is a primitive term in $\mathscr{A}_2$. Actually '$Op. Act.$' is not likely to be a primitive term in any practical theory. However, this fact is irrelevant for our present purposes. We are concerned with whether or not any attribute-correlation law-sentences such as (5.4) could be adequate connecting sentences. I think that it is easy to see that they could not.

In the first place, the correlation between optical activity and molecular dissymmetry is extremely mysterious. Any curious scientist would want the correlation explained, and would not accept it alone as providing an explanation of optical activity. In fact, scientists have developed theories to explain this correlation, see Partington (1953, Vol. 4, pp. 333–340). Naturally these explanations do not simply stipulate that this correlation is an identity, for it is not an identity. What they do is *identify* macroscopic optical activity with a disposition on the molecular level. It is then explained, within the atomic-molecular theory, why a molecule has this latter disposition if and only if it is dissymmetric. These explanations are very complicated.

In the second place, it will be noticed that this example is a correlation between a dispositional property of a whole (the optically active ESS) and a structural characteristic (dissymmetry) of this whole. This structural characteristic is not a complete structural description, but rather a geometrical aspect of certain types of structures.

Now a principal aim of a microreduction is for T_1 to explain the properties of wholes in terms of the properties of their parts, the fundamental laws governing their parts (the fundamental laws of T_1), and the boundary conditions created by the structure of the whole. We have seen an outline of such explanatory derivations in Chapter 4, Section B. Thus, the very aim of a microreduction requires that we try to *explain* in this way why molecular dissymmetry leads to optical activity. This is what the scientific theories of optical activity have done. In a very rough and intuitive way, one might say that dissymmetry is the "cause", or "part of the cause", of optical activity. But the connection between the cause and the effect needs a microreductive explanation. Dissymmetry and optical activity are merely correlated, and not identical.

Note also that the kind of microreductive explanation I have outlined above is represented by a very special kind of D-N derivation, which is really explanatory. It is a very special kind of derivation which provides a genuine part-whole explanation, i.e., an understanding of the whole in terms of its parts plus its structure. A putative microreduction that does not provide such explanations for all elements of Dom_2 and their attributes is at best an inadequate or incomplete microreduction. Such explanations are of course required for the compound elements of a theory with structured wholes.

By now it should be fairly clear that ACCS's, which are mere attribute-correlation law-sentences, are not acceptable as connecting sentences. ACCS's are mysterious, causal law-sentences that are themselves in need of explanation. If they are used as connecting sentences in **B**, then we do not explain T_2 in terms of the laws of T_1, but rather in terms of T_1 plus these mysterious correlation laws. Fortunately, actual, successful reductions do not suffer this defect. Such reductions use attribute-identities intead of ACCS's. Let us consider some examples.

A classical example of a microreduction is the kinetic theory of gases, in which macroscopic samples of gases are identified with collections of huge numbers of randomly moving molecules (or atoms). In the kinetic theory of gases, the macroscopically measured quantity, pressure, is identified with the statistically averaged change of momentum of gas molecules per unit area and time. For details one may consult, for instance, Pitzer (1953, p. 109). It is not analytic that macroscopic pressure is identical with this molecular quantity; this is rather a synthetic identity of attributes. Like all synthetic identities, this identity must be empirically justified. However, it is not explained, as correlations are. To see that it is not explained and that it is, therefore, not a mere correlation, one must be familiar with the kinetic theory of gases. In this theory, the two quantities are equated, but more important, no attempt is made to explain this equation from more fundamental laws. In fact, it appears to me that no one has even suggested trying to explain why the two quantities are connected in the way they are. When they are equated together, they are simply accepted as being identical.

One may occasionally hear it said that the pressure of a gas is simply *defined* as the average change of momentum of gas molecules per unit area and time. It may then be thought that the connection between these two quantities is analytic. However, it is certainly not analytic that gases are large collections of molecules; this was originally a very questionable empirical hypothesis. Furthermore, the language of the molecular part of the kinetic theory of gases is somewhat different and independent of the language of classical, macroscopic gas theory. The kinetic theory of gases is based on empirical connections between these two languages. It therefore is quite mistaken to assert that it is analytic that pressure is average change of momentum of gas molecules.

One must be very wary of the use of the word 'defined'. If α is discovered to be synthetically identical with β, then, after some time, scientists may begin to say that β is defined as α. This is a rather natural move to make. If $\alpha = \beta$, then this connection is noncausal, not subject to an explanation. Thus, scientists can exclude the request for a causal explanation of the connection between α and β by saying that β is defined as α. Although they may not recognize that identities are noncausal, they certainly recognize that definitions are noncausal.

This example involving pressure must not be confused with another example from the kinetic theory of gases. In this theory temperature is either correlated or identified with the statistically averaged translational kinetic energy of a molecule. Some persons maintain that this connection between temperature and a molecule's translational kinetic energy is an identity. I believe that this is the prevalent view among scientists today. Others have suggested that the connection is a correlation. This viewpoint has been expressed by Kim (1966, p. 229), who correctly points out that interpreting this connection as a correlation is *sufficient* for an extensional *derivation* of the perfect gas law. However, as I have emphasized in Chapter 2, not all D-N derivations represent explanations. Kim's reason for suggesting a correlation is therefore inconclusive.

Kim also points out that the *equation* expressing the connection between temperature and a molecule's translational kinetic energy asserts the equality only of the *values* of these quantities. This is correct, but it shows only that a mathematical equation between the measured values of two quantities does not indicate whether these quantities are identical or merely correlated. This point was discussed in Chapter 2, Section B. Therefore, when dealing with a particular case, such as this controversial temperature example, it may be quite difficult to decide whether it is an identity or a correlation. As always, it is necessary to examine the role of the example within the context of an entire theory, or a set of theories. In this respect, temperature is a very complicated example. Although I believe that it is always identified with some form of energy, it is difficult to obtain a *general* formula for the energy with which it is identified. In gases, temperature is average translational kinetic energy. In liquids and solids, temperature is identified with much more complex forms of energy.

One could go on exhibiting many more examples of both attribute-correlations and synthetic attribute-identities. For instance, the *shape* of a macroscopic crystal is identical with the shape of the molecular or ionic lattice that comprises the crystal. The change of state, *melting*, of a crystal is identical with a breakdown of the crystal lattice. On the other hand, the *property of having a certain microwave-absorption spectrum* is only correlated with having a certain molecular structure. In general, it appears that a structural characteristic, i.e., a topological or geometrical feature of a structure, can at most be merely correlated with a dispositional property. Of course, a structured whole can have other types of attributes (such as certain dispositions) in addition to having structural characteristics (such as dissymmetry).

The main results of this section are the following: An adequate micro-reduction should leave at most the fundamental law-sentences of T_1 in need of further explanation, possibly through a further microreduction of T_1. Thus, **B** should consist not merely of TICS's and ACCS's but rather of TICS's and AICS's (*attribute-identity connecting sentences*). Then all connecting sentences are identities, which are noncausal, and we can truly claim that we understand T_2 in terms of the fundamental laws of T_1.

In addition, this section has illustrated with examples the inadequacy of ACCS's, and it has exhibited examples of synthetic attribute-identities which are used as actual connecting sentences. This discussion of attribute-correlations and attribute-identities, together with that in Chapter 2, should establish quite well the inadequacy of a purely set theoretical language for many scientific purposes.

C. Summary of the Reduction Conditions

I am now in a position to state the general features of the conditions for uniform microreductions. I will start from the beginning, reviewing quickly some of the ground that has been covered.

We are given the theories T_1, T_2 formulated with the help of the sets $\mathscr{L}_1$, $\mathscr{L}_2$ of nonlogical predicates. T_1 deals with the elements in $Dom_1 = Bas_1 \cup Comp_1$, where Bas_1 is nonempty, and $Comp_1$ is nonempty in most reductions. Bas_1 is partitioned by the natural equivalence relation into homogeneous subsets, and $Comp_1$ is also partitioned into homogeneous subsets of elements according to the structures of these compound elements. Dom_2 is also partitioned by its natural equivalence relation into homogeneous subsets.

In order to have a reduction without using hidden emergent things, attributes, or laws, the language $\mathscr{L}_1$ and the theory T_1 must satisfy certain conditions. These conditions are described in some detail in Chapter 3. I will not attempt to repeat these conditions here.

Finally, the reduction involves deriving T_2 from $T_1 \cup B$, where B is a suitable set of connecting sentences. Indeed, let us now consider a *minimal* set of connecting sentences. B is minimal iff B is adequate to accomplish the reduction but any proper subset of B is inadequate to accomplish the reduction.

We saw at the end of Chapter 4, Section A, that an adequate set of connecting sentences needs at most one TICS for each primitive homogeneous thing-predicate of $\mathscr{L}_2$ which is not in $\mathscr{L}_1$. Of course, every homogeneous thing-predicate of T_2 needs somehow to be reduced. We also saw that B needs at most one ACCS reducing each primitive predicate of $\mathscr{A}_2$ which is not in $\mathscr{A}_1$. Moreover, I have argued above that ACCS's are inadequate, and that instead **B** needs AICS's.

Now each nonprimitive predicate of $\mathscr{L}_2$ is of course definable in $\mathscr{L}_2$. Therefore, if **B** contains exactly one TICS for each primitive predicate in $\mathscr{T}_2$ which is not in $\mathscr{T}_1$, and exactly one AICS for each primitive predicate in $\mathscr{A}_2$ which is not in $\mathscr{A}_1$, then any nonprimitive predicate of $\mathscr{L}_2$ can be reduced by an identity constructed from these TICS's and AICS's plus the definition of this predicate.

Putting all of this together, we see that an adequate set of connecting sentences should contain exactly one TICS for each primitive predicate in $\mathscr{T}_2$ which is not in $\mathscr{T}_1$ and exactly one AICS for each primitive predicate in $\mathscr{A}_2$ which is not in $\mathscr{A}_1$. Furthermore, no other connecting sentences are required. For suppose **B** has a connecting sentence λ which is neither a TICS nor an AICS. Then we can form λ^* by replacing all occurrences of $\mathscr{L}_2$-predicates in λ by the $\mathscr{L}_1$-predicates identical with them. Then λ^* is n-equivalent to λ. Furthermore, as we saw in Chapter 4, Section C, λ^* can be assumed to be in any adequate $\mathbf{T}_1$. Thus λ, which is then derivable from $\mathbf{T}_1$ plus the TICS's and the AICS's, is not needed as a connecting sentence. Therefore, we have a complete characterization of a minimal set of connecting sentences for microreductions, and they are all identities. Figure 2 represents the relationship between the domains of the two theories.

Now let $\mathbf{F}_i$, $\mathbf{D}_i$ be the sets of fundamental and derivative law-sentences, respectively, of $\mathbf{T}_i$ ($i = 1, 2$). Let $\mathbf{T}_2$ be reduced to $\mathbf{T}_1$ with the help of the set **B** of thing-identities and attribute-identities. In order to accomplish the reduction, each law-sentence in $\mathbf{T}_2$ must be derived from law-sentences in $\mathbf{T}_1$ plus identities from **B**. Since the law-sentences in $\mathbf{D}_2$ are explainable in $\mathbf{T}_2$, it is only necessary to derive the law-sentences of $\mathbf{F}_2$ from sentences in $\mathbf{T}_1$ plus sentences in **B**. Normally these derivations will be explanatory. However, the exact situation is not quite this simple.

Let $F_2 \in \mathbf{F}_2$. Each nonlogical predicate in F_2 is reduced by an identity sentence to a nonlogical predicate in, or defined in, $\mathscr{L}_1$. Thus, F_2 n-eq L_1, where L_1 is a law-sentence formulated in $\mathscr{L}_1$ and hence contained in an adequate $\mathbf{T}_1$. In the most usual case, L_1 is a derivative law-sentence of $\mathbf{T}_1$. Thus, there will be an explanatory derivation of L_1 within $\mathbf{T}_1$. In order to obtain a complete reductive explanation of F_2, one only needs to augment this $\mathbf{T}_1$-explanation with the appropriate identities to deduce F_2. In some special cases, L_1 may be a fundamental law-sentence of $\mathbf{T}_1$. If this occurs, one derives F_2 directly from L_1 with the help of the appropriate identities. However, this is not an explanatory derivation (i.e., causal explanation) because in this case F_2 n-eq L_1. Thus, strictly speaking, a reduction involves the following: Each law-sentence in F_2 is either (i) explanatorily derived from $F_1 \cup B$ or (ii) proved n-equivalent to a member of F_1 with the help of B. The reduction does *not* involve adding any new non-T_1 law-sentences to T_1, and so it leaves at most the fundamental law-sentences of T_1 in need of further explanation. Note that this is true in case (ii) because, if F_2 is n-equivalent to a

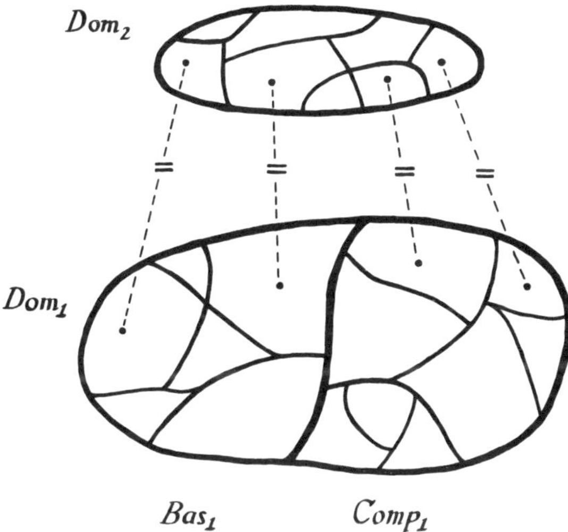

Figure 2. Diagram of the relationship between the domains. Each natural equivalence class of Dom_2 is identified with some homogeneous subset of Dom_1. These identities are asserted by TICS's. In some special cases $Comp_1$ may be empty. In some cases all identities are between the subsets of Dom_2 and subsets of $Comp_1$. In the reduction the attributes of T_2 are also identified by AICS's with attributes of T_1.

law-sentence in F_1, then F_2 states the same law as this law-sentence in F_1. Moreover, if T_1 contains some synthetic identities, they may be involved in these derivations in an obvious way. If T_2 contains some synthetic identities, then they will be derivable (but not explainable) with the help of B from corresponding identities in T_1.

With the help of B each predicate in, or defined in, $\mathscr{L}_2$, is related by an identity-sentence with a predicate in, or defined in, $\mathscr{L}_1$. Thus, the kinds and attributes referred to in T_2 are identical with kinds and attributes referred to in T_1. In other words, the ontology of T_2 is contained in the ontology of T_1. Also each law-sentence in T_2 is n-equivalent to a law-sentence in either $\mathbf{F}_1$ or in $\mathbf{D}_1$. Therefore, the set of *laws stated by* $\mathbf{T}_2$ is contained in the set of *laws stated by* $\mathbf{T}_1$.

It is important at this point to recall EE which provides a sufficient condition for two explanatory derivations to be explanatorially equivalent. EE does *not* apply to an explanatory derivation of F_2 (in $\mathbf{F}_2$) from $\mathbf{F}_1 \cup \mathbf{B}$. The reason for this lies in the fact that such a derivation contains identities and hence occurrences of *both* α and β in the antecedent of EE.[2] On the other hand, EE does have the following important application in the context of reductions. With the help of $\mathbf{B}$, and by virtue of EE, the explanatory derivations *within* $\mathbf{T}_2$ will be e-equivalent to *truncated* explanatory derivations *within* $\mathbf{T}_1$. Suppose that we have within $\mathbf{T}_2$ an explanatory derivation of a derivative law-sentence from fundamental law-sentences. Then by using $\mathbf{B}$ and repeated applications of EE, we obtain a derivation within $\mathbf{T}_1$. In general, this derivation will not be a complete explanatory derivation from fundamental law-sentences of $\mathbf{T}_1$ because in general the members of $\mathbf{F}_2$ are n-equivalent to members of $\mathbf{D}_1$. However, it is obvious that this derivation obtained in $\mathbf{T}_1$ could always be extended to a complete explanatory derivation from members of $\mathbf{F}_1$. Hence, it is natural to say that the derivation within $\mathbf{T}_2$ is e-equivalent to a *truncated* explanatory derivation within $\mathbf{T}_1$.

The above observations have demonstrated some very important consequences of my conditions for uniform microreductions. When a uniform microreduction is accomplished, then: each law-sentence in $\mathbf{T}_2$ is n-equivalent to a law-sentence in $\mathbf{T}_1$, each explanatory derivation within $\mathbf{T}_2$ is e-equivalent to a truncated explanatory derivation within $\mathbf{T}_1$, the ontology of $\mathbf{T}_2$ is contained within the ontology of $\mathbf{T}_1$, and each law-sentence in $\mathbf{F}_2$ is either explainable (i.e., understandable) in terms of $\mathbf{F}_1$ plus identities or else is seen to be n-equivalent to a law-sentence of $\mathbf{F}_1$. Moreover, the reduction does *not* involve adding any new non-$\mathbf{T}_1$ law-sentences to $\mathbf{T}_1$, and the reduction leaves at most the fundamental law-sentences of $\mathbf{T}_1$ in need of further explanation. All of these conclusions are quite natural when it is remembered that a microreduction is intended to show that the phenomena accounted for by $\mathbf{T}_2$ are the same as some of the phenomena in the sphere of $\mathbf{T}_1$. In this regard a very special, rare possibility should be mentioned. Suppose that Dom_2 is identical with Bas_1, that the set of attributes of $\mathbf{T}_2$ is identical with the set of attributes of $\mathbf{T}_1$, and that there is a one-to-one correspondence between $\mathbf{F}_2$ and $\mathbf{F}_1$ such that corresponding law-sentences are n-equivalent. Then we can say that $\mathbf{T}_1$ and $\mathbf{T}_2$ are *equivalent theories*.

A terminological comment is in order here. According to my conditions for microreductions, **B** consists only of identities. But an identity-sentence is not a law-sentence. Therefore, the prevalent term, "bridge law", for connecting sentences, is obsolete and incorrect. A microreduction does not, and should not, use any law-sentences as connecting sentences. It is to be hoped that the incorrect term, "bridge law", will fade into disuse.

Of course, an attribute-correlation is not an attribute-identity, and it cannot be changed into an attribute-identity. Thus, the requirement that the connecting sentences be identities is a very strong requirement which will make microreductions difficult to achieve. In practice, I believe that scientists have usually found suitable identities in their successful reductions. Naturally, there can be cases of putative identities that are very difficult to justify as identities. Many kinds of identity claims, even concerning particular objects, are difficult to justify. The justification of the identity connecting sentences will be accomplished within the entire framework of a successful reduction. More will be said about this in the next section.

D. SOME POSSIBLE OBJECTIONS AND PROBLEMS

I believe that I have convincingly demonstrated both the need for, and the possibility of, synthetic attribute-identities. Nevertheless, it is still possible to raise certain objections and problems concerning them. Many possible objections have already been answered in previous discussions. In this section I will briefly consider a few other objections which have been, or might be, made. I will also discuss a general problem and show that it is not as serious as it might at first appear to be.

1. Let α, β be two attribute-predicates which are synthetically co-extensional. I say that α and β are identical if and only if this co-extensionality is noncausal. If the co-extensionality is a law, then I say that we have a nomological co-extensionality, or an attribute-correlation law. We thus have a distinction between attribute-correlation law-sentences and attribute-identities. Some have suggested that this is really nothing more than the distinction between *accidental* co-extensionalities and *nomological* ('lawlike') co-extensionalities.[3] I believe that the extensive discussion in Chapter 2 is sufficient to show that this suggestion is incorrect. Moreover, the arguments in the present chapter show that a nomological co-extensionality is not acceptable as a connecting sentence

in an adequate reduction. For instance, the correlation between optical activity and molecular dissymmetry is certainly a well-confirmed law which has also been explained with some success. Yet, it is not an acceptable connecting sentence, and it is certainly not an attribute-identity. It should therefore be quite clear that the distinction between attribute-correlations and attribute-identities is not the same as the distinction between accidental and nomological co-extensionalities.

Nevertheless, it might still be suggested that what I have called 'identities' are just very special types of correlation laws which are somehow especially suited for use as connecting sentences in reductions. But in order for such special correlations to serve as connecting sentences, they would have to be noncausal, and hence by the arguments in Chapter 2 they would not be laws. Moreover, such special correlations would not identify the kinds and attributes of T_2 with the kinds and attributes of T_1. Hence, they could not really be adequate connecting sentences. The suggestion of such special correlation laws faces many other obvious difficulties, but it would be superfluous to mention them here.

2. In view of the arguments of Section B of this Chapter, it is highly unlikely that anyone would object that ACCS's are sufficient for micro-reductions. Yet, such an opinion concerning reductions was still quite prevalent rather recently. It has been claimed that, for scientific purposes, there is really no significant difference in whether **B** contains identities or attribute-correlations. For instance, in discussing the psycho-physical identity theory, Brandt and Kim (1967, p. 534) consider the argument to the effect that the identity theory can be defended on the basis that it can provide predictions and explanations of psychological laws. They then reply, "The trouble with this argument is that the deduction of the psychological laws from neurological theory would proceed just as well if the correlation laws were regarded just as correlation laws, and not as identities. Anyone who uses this argument for IT [the identity theory] must show that the deduction succeeds *only if* identity statements are used in the deduction, and not correlation laws."

Their reply is concerned with IT, but if it were correct, it would apply to microreductions also. However, their argument seems to be based on two incorrect presuppositions. First, they imply that it is simply a matter of *interpretation* whether a connection between two predicates is a correlation law or an identity. But this is certainly not the case. An attribute-correlation law cannot, by fiat, be changed into an attribute-identity, nor vice versa. Second, their reply makes a significant move from *explanation* to *deduction*. They fail to recognize that an adequate explanation is not represented by just *any* D-N derivation. In particular, an *explanatory* derivation from T_1 plus identities will in general be quite different from a D-N derivation from T_1 plus attribute-correlation law-sentences. Therefore, even if a law-sentence can be D-N-derived from T_1 plus attribute-correlation law-sentences, this derivation is not necessarily an explanatory derivation. Moreover, even if it is an explanatory derivation, it is a derivation which uses new law-sentences in addition to those of T_1. Therefore, it cannot represent a *reduction* to T_1.

3. At several places I have mentioned the traditional doubts about the possibility of synthetic attribute-identities. It has been shown that these doubts were unjustified. In addition, in recent years increasing numbers of philosophers have been willing to accept the existence of synthetic attribute-identities. In spite of this fact, it has been claimed that synthetic attribute-identities have a logical structure which is unsuited for their use as connecting sentences in reductions. For instance, Jaegwon Kim at one time made such a claim about contingent property-identities, and thus presumably also about synthetic attribute-identities. Kim reasoned in the following manner. After considering some examples of contingent property-identities, he generalized to a conjecture, for which he admitted he had no proof. In Kim (1966, p. 233) it was conjectured that ". . . all contingent statements of property identity contain, essentially, some expression that refers to a particular or individual". He also thought that, if there were a statement of identity between a kind of pain and a kind of brain state, such a statement would probably not refer essentially to a particular. He concluded that such an identity would be unlikely to be a contingent and nonanalytic identity of properties. It should be noted that Kim himself qualified this conclusion and did not appear entirely confident of it, and his recent work indicates that he has now perhaps rejected it. Nevertheless, if this conjecture were correct, it would imply that there could be no general microreductive connecting sentences which are synthetic attribute-identities.

Fortunately, the conjecture is incorrect. One of Kim's own examples of a contingent property-identity was "Black is the color of ravens." This statement has no reference to a particular and is hence a counterexample to the conjecture. Apparently Kim believed that it was not a counter-example because it refers to a particular species of bird. We can grant this for the sake of the argument. Then all we need do is consider the counterexample: *black* is identical with *reflects no visible radiation when illuminated*. In this example, '*black*' refers to an objective color which may be possessed by certain opaque objects. This identity refers to no particulars, and it is synthetic. It is certainly not analytic that *black* is in any way connected with reflection of radiation. For instance, scientists might have discovered empirically that black objects are objects that emit certain kinds of particles. There could be different particles for red, green, black, etc. Furthermore, if this counterexample is not convincing, several others have already been mentioned in Section B of this chapter. Thus, the conjecture is false, and it should not be considered a barrier to the existence of attribute-identities suitable for use as connecting sentences in microreductions.

4. Saul Kripke defines a '*rigid designator*' to be a term which designates the same object in all possible worlds in which this one object exists. The notion of 'possible world' is that which is used in some systems of modal logic, but it also is used rather loosely to refer to subjunctive and counterfactual conditions. Kripke claims that if we have a true identity between rigid designators, then this identity is also *necessarily true*, which means that it is true in all possible worlds. In Kripke (1971, p. 150) he also says, informally, that a statement is necessary if it is true, and if it could not have been otherwise. Among other things, Kripke believes that certain terms of scientific languages are rigid designators, and consequently he believes that certain scientific identities are necessarily true if they are true. For instance, he argues (Ibid., p. 158–160) that it is necessary that heat is the motion of molecules. From a scientific point-of-view, this is not a very well formulated statement of identity, but let us not be distracted by this fact. Instead, I wish to comment very briefly on two possible relationships between Kripke's views and those defended in this book.

First, if Kripke's views are correct, then it would appear that most, if not all, true TICS's and AICS's would be necessarily true. It might then appear (as has been suggested to me) that there could be no synthetic connecting sentences, since it might appear that there could be no necessary synthetic sentences. However, Kripke uses the term 'necessary' in a rather specialized metaphysical sense (Ibid., p. 150), and I have used the terms 'analytic' and 'synthetic' in a linguistic, conventionalistic sense. It is not at all obvious how these two systems of terminology are related. Therefore, even if we assume that Kripke's views are correct, it is not at all clear that they conflict with my claim that there can be synthetic TICS's and AICS's. I will not attempt to resolve this obscure issue here since I do not believe that it is a significant issue. In fact, I am not convinced that Kripke has shown that there is any significant sense in which it is necessarily true that heat is motion of molecules.[4]

Second, it has been suggested to me by several persons that my thesis that thing- and attribute-identities are noncausal is somehow equivalent to, or follows from, Kripke's thesis that true identities are necessarily true. This suggestion was not intended as an objection to my thesis, but rather as a possible justification or reinterpretation of it. Nevertheless, I am not convinced that this suggestion is correct.

As I said above, I am not convinced that it has been shown that it is necessarily true (in any significant sense) that heat is motion of molecules. But even if it is assumed that Kripke's thesis is correct, it does not have any direct connection with my thesis that thing- and attribute-identities are noncausal. Kripke's arguments concerning the necessity of true identities do not involve the notions of causation or causal explanation. My arguments that thing- and attribute-identities are noncausal depend on the semantics of these identities plus certain aspects of causation and causal explanation. My arguments do not utilize any concept of metaphysical necessity. In order to obtain my results from Kripke's, it would at the very least be necessary to supplement his views with additional assumptions about causation and causal explanation. Moreover, my results can be obtained independently of his because my arguments do not utilize any notion of metaphysical necessity. I do not claim absolutely that there are no relationships between his views and mine, but it should be quite clear that his thesis does not imply mine.

5. Since the preceding discussion has taken us into the realm of modalities, it is appropriate here to mention another possible suggestion. In the present chapter and in Chapter 2 I have presented several reasons why set theoretical languages are inadequate for some scientific purposes. I have defended instead a language with predicates referring to kinds and to attributes. It might be suggested that this move to kinds and attributes is unnecessary, and that instead we should use a modalized set theoretical language. Such a language would use set theoretical interpretations of its predicates, and in addition it would contain a modal operator for 'noncausal'. If this operator is placed in front of a co-extensionality, then this co-extensionality would play a theoretical role

in some ways analogous to that played by a thing- or attribute-identity. 'Not noncausal' would then be placed in front of any law-sentence. (I assume that accidental co-extensionalities would not be used in a dynamic theory).

I have no doubt that such a language could be constructed. However, it would be extremely ad hoc, and its construction would be basically an empty, formal exercise. Such a language would have no natural reason for the existence of the modal operator for 'noncausal'. On the other hand, if we include kinds and attributes in our ontology, then it can be argued from the semantics of identities plus aspects of causation and causal explanation that thing- and attribute-identities are noncausal. Then the notions of n-equivalence and e-equivalence have a natural, intuitive interpretation, and reductions include reduction of the total ontology of a theory (including kinds and attributes). All of these intuitive interpretations and results would be lost or extremely distorted in a set theoretical language with a modal operator for 'noncausal'. It seems clear that such a language does not offer a viable alternative to the analysis presented in this book.

6. Let us now accept TICS's and AICS's as the proper kinds of connecting sentences for microreductions. This makes microreductions difficult to achieve. It might be thought that this creates a very difficult practical problem: how can we ever be sure that a genuine microreduction has been accomplished? It is not sufficient merely to derive the fundamental law-sentences of T_2; they must be adequately derived from T_1 plus identities, and it may be difficult in practice to establish whether a connecting sentence is an attribute-identity or a mere attribute-correlation law-sentence.[5] I have already conceded that this can be a difficult matter in certain cases. However, I do not think that it is as serious a problem as it might at first appear to be.

As scientists work on a progressing microreductive research program they will gradually discover and confirm increasing numbers of co-extensionalities. Most of these will probably turn out to be attribute-correlation laws. I have already pointed out that it appears that all co-extensionalities between structural characteristics and dispositional properties are attribute-correlations. These, and other co-extensionalities will be explained as the microreductive program proceeds. Any co-extensionality which is explained is of course an attribute-correlation and not an identity. Eventually there will remain a residual set of connecting sentences which will resist continued efforts at their explanation. It will be natural to hypothesize that this residual set is a identities. There will be no way to demonstrate with certainty that ... residual connecting sentence is a genuine identity and hence noncausal. As time goes on some of them may in fact eventually be explained. Suppose now that the reduction is finally formulated in what appears to be the most concise possible form, using what appears to be an absolutely minimal set of connecting sentences. Also suppose that all of the explanatory derivations used in the reduction seem quite adequate, and appear to give a systematic account of all of the fundamental laws of T_2. Then, unless there are some special reasons to the contrary, there will be good justification that the connecting sentences are identities. Of course, any empirical hypothesis is subject to later revision.

Even under the circumstances described, it might still be claimed that some or all of the connecting sentences are mere correlation law-sentences. Such a claim would be extremely peculiar in the case of the thing-identities, and therefore it would almost certainly not be made in this case. So suppose that it is claimed that the reduction uses thing-identities, but mere attribute-correlations. If the person making this claim concedes that these alleged correlations could be explained by T_1 plus some genuine identities, then the burden is on him to do so. If he succeeds, then he has in effect constructed a genuine microreduction, and he has shown that our original attempt at a microreduction was in fact defective.

On the other hand, he might claim that these alleged correlations are some kind of absolutely fundamental laws which, as a matter of *fact*, could never be explained. This implies that the microreduction cannot be accomplished. In addition, it also implies that the world is governed by some extremely peculiar, fundamental attribute-correlation laws. I believe that such an hypothesis will, in general, appear less plausible than hypothesizing certain appropriate attribute-identities in the kind of situation considered here.

E. REASONABLE MODIFICATIONS

When seeking an explanation of E, one often ends up explaining some E^* which is approximately the same as E. I illustrated and discussed this well-known fact in Chapter 2, and it is relevant to microreductions. In practice the scientist usually does not derive a previously given T_2, but rather a theory T_2^* which is *approximately* the same as T_2.

Usually, T_2* is a more detailed, more complex, and more precise theory than T_2.

In addition to being more precise than T_2, T_2* sometimes differs from T_2 in another way. Occasionally the language $\mathscr{L}_2$ of T_2 undergoes some modifications for use as the language of T_2*. Some, e.g., Feyerabend (1962), have referred to such modifications as "meaning changes", and they have considered them to be extremely important.

Finally, if scientists begin with a T_1 and a T_2 and attempt a reduction, they will usually also develop and perhaps change T_1 in the process. If the reduction is ever completed, they might end up with a T_2* reduced to a T_1*, where both of these new theories are related to the original theories by approximations and modifications.

Of course, the conditions for microreductions which I have developed here are conditions to be satisfied by *completed* microreductions. These conditions should therefore be satisfied by T_2* and T_1*. Nevertheless, if the approximations and modifications are *reasonable* changes of the original theories, then I think that it is also fair to say that T_2 has been reduced to T_1. In order to throw more light on this issue, I will briefly indicate some kinds of modifications which are both common and, it seems to me, reasonable. It will then also be possible to clear up an old problem about trivial microreductions.

Let us first consider some possible modifications of $\mathscr{L}_2$. For instance, it occasionally happens that a T_2 classifies things in an infelicitous manner with respect to T_1. In fact, T_2 may even overlook an important kind of thing. For instance, as T_1 is developed it may be discovered that a certain kind of compound element of Dom_1 can be produced which was never encountered or conceived of before, and hence not named by any predicate defined in $\mathscr{T}_2$. In such a case it is natural and reasonable to add an appropriate name for this kind of thing to $\mathscr{T}_2$. Often its $\mathscr{L}_1$-name is adequate for this purpose.

More serious kinds of modifications are sometimes in order. Occasionally it might happen that a predicate 'W' of $\mathscr{L}_2$ is believed to be a homogeneous thing-predicate. Then later it may be discovered that it is really not homogeneous, and that it is impossible to find a true TICS reducing W's to a homogeneous class of Dom_1. For example, air was once believed to be a chemical element, and hence believed homogeneous. Then it was later discovered to be a *mixture* of various substances. It was also discovered that air is not homogeneous. For instance, a macroscopic sample of air from Los Angeles has different properties from a sample of air from the Amazonian jungle. Therefore, 'air' is no longer considered a homogeneous thing-predicate which needs to be reduced by a TICS. Instead, 'air' is understood as a heterogeneous thing-predicate to be defined as a mixture in terms of homogeneous thing-predicates.

Needless to say, if we modify our classification of Dom_2 to obtain a suitable accord with Dom_1, we do not necessarily end all possible doubts about the ultimate correctness of our classification scheme. We only use T_1 as a higher authority than T_2. There can still be doubts about the correctness of T_1. But if T_1 becomes well-confirmed, then we would certainly be justified in using it as the higher authority. If it is in turn microreduced to an even lower level theory, then this new lower theory might eventually become an even more powerful authority than T_1 and require some modifications in T_1 (and then perhaps indirectly in T_2).

Notice that, as we move from the basic elements of Dom_1 to the compound elements of Dom_1, there is not only a proliferation of kinds of things, but also of kinds of attributes. One of the beautiful aspects of successful microreductions is that they provide us with a more unified, and in some ways more simple, view of the world. In fact, it is usually the case that the set of classifying attributes for classifying elements of Bas_1 is a smaller set of classifying attributes than is used in T_2 for classifying the elements of Dom_2.

I believe that most scientists, and probably most philosophers, will agree that it is reasonable to say that we have reduced T_2 to T_1 if this reduction only requires some small modifications of the classification scheme of T_2. On the other hand, some modifications of $\mathscr{L}_1$ would probably be controversial. For instance, some modifications of $\mathscr{L}_1$, if they are not suitably restricted, can lead to *trivial* microreductions. One familiar kind of trivial microreduction is the following, derived from Oppenheim and Putnam (1958, p. 10). I will use primitive terms to simplify the example, but this is not necessary. Let 'A' be a primitive attribute-predicate of $\mathscr{L}_2$, and let 'W' be a primitive basic homogeneous thing-predicate of $\mathscr{L}_2$, and suppose that

$$Wx \to Ax, \tag{5.5}$$

is in T_2. We want to explain (5.5) by T_1, but we have no AICS for 'A', so we must get one somehow. We use the part-relation, 'Pt', which can be defined in $\mathscr{L}_1$. This relation satisfies the condition $Pt\, yx$ if and only if x is a compound element of Dom_1 and y is a member of the domain of x.

Now, in order to get an AICS for 'A', we add a new predicate, 'A''', to $\mathscr{A}_1$, and add to **B**

$$Ax \leftrightarrow (y)(Pt\ yx \rightarrow A'y). \tag{5.6}$$

In the Oppenheim-Putnam example, 'A' is 'transparent,' and 'A''' is the fabricated property, 'Tran,' the property of being an atom of a transparent substance. Thus, for (5.6) they have a sentence like: x is transparent if and only if every atom of x is *Tran*. This seems very suspicious, but nevertheless we proceed to add to **B**

$$Wx \rightarrow (y)(Pt\ yx \rightarrow A'y). \tag{5.7}$$

Clearly, (5.5) follows from (5.6) and (5.7), and we have our trivial micro-reduction.

In order to avoid such trivial reductions, Oppenheim and Putnam assume that the world has various "levels", i.e., elementary particles, atoms, molecules, etc. They further assume that at present there is a fixed set of theoretical predicates appropriate to each level. They then require that reductions only be performed in terms of these fixed sets of predicates. This corresponds to fixing $\mathscr{L}_1$, $\mathscr{L}_2$ at some time and then never allowing any modifications of them. This is an impractically rigid restriction on reductions. What, then, are we to do?

First of all, observe that (5.7) is not allowed in **B** according to our previous characterization of **B**. However, it can be eliminated from **B** by adding

$$\phi x \rightarrow (y)(Pt\ yx \rightarrow A'y) \tag{5.8}$$

to $\mathbf{T}_1$, where ϕ is the $\mathscr{L}_1$-predicate with the same denotation as 'W'. Thus, we have now modified both $\mathscr{L}_1$ (by adding 'A''' to it) and $\mathbf{T}_1$ (by adding (5.8) to it). Furthermore, we still have the suspicious-looking connecting sentence (5.6).

However, on further consideration one can see that (5.6) and (5.8) would not seem suspicious if 'A''' had been an original term of $\mathscr{A}_1$ with a clear interpretation. In fact, it is not hard to imagine situations in which sentences like (5.6) and (5.8) are true. It *is* clear that we would be suspicious of them if 'A''' were added to $\mathscr{A}_1$ and 'A''' has no use other than its role in these sentences. In such a case these additions would really seem *ad hoc*.

On the other hand, there could be situations in which adding 'A''' would be adding an important new theoretical concept to $\mathscr{A}_1$. But certain conditions would have to be met. For instance, if (5.8) is true, then we would usually expect that other kinds of compound elements in addition to ϕ's would be such that all of their parts have the property A'. Thus, there would be wholes other than just W's which have the property A. Thus, adding 'A''' to $\mathscr{A}_1$ would normally involve adding many sentences like (5.8) to $\mathbf{T}_1$. More generally, if we added 'A''' to $\mathscr{A}_1$, then we would want to determine somehow what $Ext_{A'}$ is. One way of getting information about this extension would be by finding the various kinds of compound elements of Dom_1 which have A. This would involve finding all the appropriate sentences like (5.8). Finally, if we were still not satisfied that (5.6) is an AICS, but think that it is an ACCS, then we might try to explain it with the help of some other proposed AICS reducing A and involving A'. If this were done, we would not use (5.6) as a connecting sentence, but we would continue to accept it as a true sentence. Thus we still would allow the addition of 'A''' to $\mathscr{A}_1$, and we would no longer believe that we have only accomplished a trivial microreduction. Indeed, we might then also find other laws of $\mathbf{T}_1$ involving 'A''', and thus we would have accomplished an important and reasonable advance in $\mathbf{T}_1$.

In fact, such advances have occasionally taken place. When John Dalton was developing his atomic theory, he wanted, among other things, to explain the various combining weights of substances. For instance, why do 35.457 g of chlorine react exactly with 1.008 g of hydrogen? Dalton, as is well known, invented the notions of atomic and molecular weights in order to explain the existence of the combining weights. His job was much more involved than ours, for he had to explain many quantitative relations among many substances. Thus, in rough analogy, he had many A's and A''s with which to deal. He did not complete the task himself, but eventually the work of many chemists led to the development of a consistent system of atomic weights. If these chemists had been restricted in the way Oppenheim and Putnam recommend, they would not have been allowed to introduce the new concept of atomic weight. It should now be clear that the Oppenheim-Putnam requirement of *fixing* $\mathscr{L}_1$ and $\mathscr{L}_2$ prior to a microreduction is impractical.

Needless to say, I do not have general rules for guaranteeing the exclusion of all possible kinds of trivial microreductions. However, we should be allowed to expand $\mathscr{A}_1$, if necessary, provided the expansion is done consistently, the denotations of the new predicates are determined, and all related appropriate modifications are made in $\mathbf{T}_1$. Using a $\mathbf{B}$ which is a minimal set of TICS's and AICS's should provide a useful constraint in such expansions of $\mathscr{A}_1$.

It has not been my intention to give an exhaustive discussion of different kinds of reasonable modifications. However, I believe that the examples above should provide a good idea of some kinds of ways in which a scientist might expect to have to modify his theories in the course of constructing a reduction. More importantly, these examples illustrate how our conditions for completed uniform microreductions can be a heuristic guide in deciding how one can reasonably modify his theories.

NOTES

[1] Putnam also says that the nonexistence of unicorns might be a law of nature. He says that things become still more complicated if this is the case. This is indeed true for his discussion of this *example* because here he distinguishes the nonreducing from the reducing co-extensionality by means of the accidental/nomological distinction. In this chapter I argue that all connecting sentences are thing- or attribute-identities, and I have previously argued that these identities are noncausal and are not law-sentences. Therefore, it is irrelevant to my analysis of reduction whether the co-extensionality involving 'Unicorn' is accidental or nomological.

It should be remarked that I agree with the spirit of Putnam (1969), namely, that reductions involve identities of things and properties. Moreover, Putnam's discussion makes it quite clear that he is aware of the difficulties involved in adequately explicating the nature of these identities. He suggests alternative explications, but it is not clear to me exactly what his final position is.

[2] For this reason EE does *not* imply "explanatory symmetry" of reductive explanations, as has been (apparently) suggested in Enç (1976, pp. 287–291).

[3] In Putnam (1969, pp. 237, 241, 253) it is suggested that nomological co-extensionality might be used as a criterion for property identity, although Putnam actually defends a different proposal in this article. Putnam's suggestion is partially defended and partially qualified in Malinas (1973). Malinas does not specifically state the suggestion mentioned in my main text, but he does say that a sufficient condition for property identity is that two predicates be related by a nomological co-extensionality used as a connecting sentence in a reduction.

[4] When I wrote Causey (1972a) I was unaware of Kripke's thesis about the necessity of true identities. In Causey (1972a) I used the term 'contingent identity' rather than 'synthetic identity'. This was unfortunate, for all that was meant by 'contingent identity' was what I now mean by 'synthetic identity'. However, because I used the term 'contingent identity', some readers concluded that there was a serious conflict between my position and Kripke's. As the discussion in the present chapter shows, there may be no conflict at all. Indeed, some philosophers have suggested that Kripke's thesis and mine are closely related. I do not share this view.

It will be noticed that the terminology in Causey (1972a) also differs in other respects from that used in my more recent articles and in this book.

[5] In their criticism of Causey (1972a), Ager *et al.* (1974, p. 131) express fear that Causey ". . . has assigned himself the Promethean task of deriving an '=' from an '≡'." This is a misunderstanding. I have nowhere suggested any such derivations, and they would be impossible. If P, Q are nomologically co-extensional attributes, then P and Q are causally related and distinct. A true identity between P and Q cannot be derived from a nomological co-extensionality between P and Q. For a reply to this and other criticisms by Ager *et al.*, see Causey (1976a).

CHAPTER 6

UNIFIED THEORIES AND UNIFIED SCIENCE

Section A of this chapter contains a brief review of some previous conceptions of the unity of science. In Section B I present and discuss some conditions which must be satisfied by a dynamic theory in order for that theory to be unified. In Section C an argument is presented to justify the use of microreductions as a unifying procedure. The principal conclusion is: If T_2 can be microreduced to T_1, then the only satisfactory way to unify these two theories is by a microreduction to T_1, or to a theory which is a modified formulation of T_1. This result provides a rational foundation for a program for the unification of science based on successive microreductions of various theories.

The material in Sections B and C represents a development of earlier work presented in Causey (1976b). The definitions and conditions presented here differ in important respects from those in the work cited. The arguments have also been extensively revised and expanded.

A. MICROREDUCTIONS AND UNIFIED SCIENCE

1. As I remarked in the very beginning of this book, there has long been interest in the possibility of a unified science which would provide a general understanding of the various classes of natural and social phenomena. In the Twentieth Century this interest was first very strongly promoted by the Logical Positivists. Eventually a series of international meetings on the unity of science was organized, and there was founded an Institute for the Unity of Science (which recently dissolved itself). A board of editors was established to supervise the publication of the *International Encyclopedia of Unified Science*. Volume 1, Number 1, Neurath, *et al.* (1938), contains an article by Rudolf Carnap titled 'Logical Foundations of the Unity of Science'.

In this early discussion of the unity of science, Carnap distinguishes unity of laws from what he calls "unity of the language of science". He claims (p. 61) that unity of language is a necessary preliminary condition for the unity of laws. In the spirit of Logical Positivism, he argues that all branches of science should "reduce" their terms to what he calls the "physical thing-language". The type of "reduction" Carnap proposes is quite different from the procedure of microreduction which I have analyzed in previous chapters. Furthermore, Carnap's physical thing-language is constructed according to certain epistemological requirements; it should not be confused with the language of any particular theory, nor with my previous discussions of thing-predicates. Carnap realized the importance of trying to obtain unity of laws. He also recognized the logical point that some kind of unity of language was necessary in order to achieve unity of laws. However, in my discussions below, I will not make use of a concept of unity of language in anything like Carnap's sense. Our understanding of the requirements for the unification of science has undergone extensive change and development since Carnap's important, early work on the subject.[1]

2. In 'Unity of Science as a Working Hypothesis', Oppenheim and Putnam (1958) present a program for the unification of science based on successive microreductions. I will review some of the more important features of their program because it has become rather influential, and it is related to the present work.

Oppenheim and Putnam begin by distinguishing unity of language from unity of laws. They recognize that the mere conjunction of the laws of different branches of science will probably not be an acceptable method of unification. Unity of laws is to be obtained by the reduction of all laws of science to the laws of one scientific discipline. In addition, they write, ". . . Unity of Science in the strongest sense is realized if the laws of science are not only reduced to the laws of some one discipline, but the laws of that discipline are in some intuitive sense 'unified' or 'connected'. It is difficult to see how this last requirement can be made precise; and it will not be imposed here" (p. 4). I will return to this point below.

Although they do not have conditions for the unity of a theory, Oppenheim and Putnam believe that the appropriate method of unification is through successive microreductions. I will not describe their conditions for microreductions. However, it should be remarked that their conditions are considerably different from those I have elaborated in earlier chapters. Nevertheless, their conditions for microreductions do require a part-whole relationship between the elements of the domains of the theories involved. Ignoring certain technical qualifications, the Oppenheim-Putnam unification program can be outlined as follows.

Assume an idealized view of the structure of science in which science consists of several branches, each branch consisting of a theory (or theories) about a certain domain of objects. In addition, assume that these domains of objects (and hence their corresponding theories) are linearly ordered so as to form a hierarchy of levels. In this ordering, let there be a lowest level consisting of a domain of indecomposable elements. Assume that any element in the domain of any higher level is a structure whose parts are elements of the next lower level. In particular, Oppenheim and Putnam mention six levels. The lowest level consists of elementary particles, above this is the level of atoms (in the sense of the current atomic theory), then the level of molecules, the level of living cells, the level of multicellular living things, and finally the level of social groups. Oppenheim and Putnam realize that this is an oversimplified description of the structure of science, and they qualify their description in some ways. However, the basic goal of their unification program is the micro-reduction of the theory (or theories) of each higher branch to those of the next lower branch, and hence the successive microreduction of all branches to the lowest branch dealing with elementary particles. They recommend that it be a scientific working hypothesis that this goal can be achieved.

3. Many questions can be raised about the Oppenheim-Putnam unification program. Among the more important are the following: (i) What conditions must be satisfied in order for a theory to be unified? (ii) What conditions must be satisfied in order to achieve an adequate microreduction? (iii) Why should the unification program be based on successive microreductions? (iv) What reasons are there for believing that the proper empirical relationships will exist between the various branches of science?

Oppenheim and Putnam devote a large portion of their article to question (iv). They review many scientific developments in order to provide evidence that suitable microreductions have been, or someday could be, performed. On the other hand, as the quotation above shows, they do not even attempt to answer question (i).

In connection with question (ii), Oppenheim and Putnam do present a set of very general conditions for microreductions. Their reduction conditions are much different from mine, and, although I will not argue the point here, I believe that they are inadequate. Moreover, Oppenheim and Putnam recognize that their reduction conditions do not exclude trivial microreductions of the kind discussed in Chapter 5, Section E.

Thus, they impose the ad hoc requirement that the nonlogical predicates of each branch be fixed, and that all reductions be performed within this linguistic restriction. As I argued in Section E of Chapter 5, this is an unreasonable and impractical restriction on a unification program. Indeed, Oppenheim and Putnam (p. 13) apparently are willing to allow exceptions to their own restriction. There are no absolute guarantees against the possibility of trivial microreductions. However, I do not believe that they will be a serious threat if microreductions are performed according to the conditions described in previous chapters of this book.

Oppenheim and Putnam present some general discussion related to question (iii). Primarily, they mention certain desirable consequences of unification by microreductions. Unfortunately, they do not present any very detailed argument to show that unification by means of micro-reductions is theoretically superior, rather than merely practically preferable, to other possible methods of constructing a unified science. The absence of such an argument is not surprising in view of the fact that they do not attempt to answer question (i).

Oppenheim and Putnam do make an interesting observation which is suggestive in this context. They point out that our current theories of the evolution of the universe indicate that there were past times during which elements of level n existed, while no elements of higher levels existed. Presumably, as time passed, elements of higher levels evolved from elements of lower levels. Let us also suppose that it is possible to give causal explanations of such evolutionary processes. For instance, if a certain type of structured whole has evolved from certain types of parts, then we assume that this evolutionary process can be causally explained. Oppenheim and Putnam suggest that it is then natural to suppose that the theory of such wholes could be microreduced to the theory of the parts of such wholes. They believe that this reasoning provides some factual support for the feasibility of their working hypothesis. Of course, this reasoning does not show that the micro-reduction can be performed. As Oppenheim and Putnam recognize, it is logically possible that new, irreducible, emergent properties of wholes might evolve with these wholes. Moreover, even if the evolutionary assumptions are granted, it is still not obvious that microreduction provides the only method, or the optimum method, of unifying the theories involved.

One of the most severe critiques of the Oppenheim-Putnam unification program is found in Schlesinger (1963, Chapter 2). Schlesinger not only criticizes particular aspects of the Oppenheim-Putnam program, but he also makes some bold claims about the nature of explanation. Suppose that T_2 is a theory about a domain of wholes and T_1 is a theory about a domain whose elements are the various kinds of parts of the wholes governed by T_2. Most scientists and philosophers would probably say that the proper way to attempt to unify T_1 and T_2 is by a microreduction. Schlesinger claims that this is an unjustifiable partiality resulting from a tendency to equate physical order with logical order. It is therefore interesting to note his interpretation of the evolutionary consideration mentioned above.

Let T_2 be a theory of living cells and let T_1 be a theory of molecules. Suppose that somehow T_1 could be explained by T_2; Schlesinger calls this process a "macroreduction". Also suppose that cells evolved from molecules. Then it would appear that there was a past time during which the behavior of molecules was indeterminate. Or else during this past time the behavior of molecules was determined by laws governing the behavior of cells which did not yet exist. The former conclusion seems absurd, and the latter is very mysterious.

Schlesinger admits that these considerations have suggestive power which motivates the use of microreductive explanations. Nevertheless, he believes that this suggestive power is just another example of the tendency to confuse causal and logical relationships. "There is no reason why propositions about the behavior of non-existent entities (and if you like of entities that will never exist) should not, in the context of a certain theory, logically determine or imply propositions about the behavior of existing entities", Schlesinger (1963, p. 51). Schlesinger may be correct in claiming that propositions about the behavior of existent entities could be derived from certain propositions about nonexistent entities. But there is no reason to suppose that such derivations would represent *causal explanations*, nor is it obvious that such a "theory" could be a unified theory. Nevertheless, I accept Schlesinger's remark as a challenge to justify the use of microreductions as a unifying procedure. Such a justification must be based, at least in part, on conditions which must be satisfied in order for a theory to be unified. Fortunately, it is not necessary to have a complete set of sufficient conditions for a theory to be unified. A satisfactory justification can be based on an appropriate set of conditions which are necessary for a theory to be unified. This set of necessary conditions is presented in the next section.

B. UNIFIED THEORIES

1. Consider a theory **T** with the domain, *Dom*. The language of **T** describes kinds of objects in *Dom*, as well as attributes of these objects. As in earlier chapters, this language can use symbols of set theory and mathematics (its 'logical symbols'). The set of nonlogical predicates of this language is $\mathscr{L}$. $\mathscr{L}$ is the union of two disjoint subsets, $\mathscr{T}$ and $\mathscr{A}$, where $\mathscr{T}$ is the set of thing-predicates and $\mathscr{A}$ is the set of attribute-predicates. As usual, a thing-predicate denotes a kind of element in *Dom*, and an attribute-predicate denotes an attribute applying to elements of *Dom*. Some predicates of $\mathscr{L}$ may be definable in terms of others, but no *definite descriptions* of particulars, kinds, or attributes are used in the language of **T**.

T will, of course, contain law-sentences, which state laws about the attributes and behavior of the various kinds of things in *Dom*. **T** may also contain identities, which can be thing-identities between two thing-predicates or attribute-identities between two attribute-predicates. We recall that identities are noncausal sentences; they are not subject to causal explanation, although they may be included in the explanans of causal explanations. The two predicates occurring in an identity are co-extensional, but the identity is not a law-sentence and the co-extensionality is not a nomological co-extensionality.

Two law-sentences, L_1, L_2, formulated in $\mathscr{L}$, may state the same law and hence be nomologically-equivalent. In general, by condition NE, L_1 n-eq L_2 if and only if there is a set **N** of true noncausal sentences of $\mathscr{L}$ such that $L_1 \leftrightarrow L_2$ is derivable from **N**. However, in this chapter it is supposed that **T** is a dynamic theory. Some general features of such theories were described in Chapter 2, Section C. The languages of such theories are relatively simple, and it appears likely that any noncausal sentence in such a language is derivable from analytic sentences plus synthetic identities. Moreover, the explanatory derivations in dynamic theories have explanans which may contain fundamental law-sentences, identities, and analytic sentences. In effect, this presupposes that the noncausal sentences of these theories are derivable from analytic sentences plus synthetic identities. But analytic sentences are always removable from the premises of a derivation. Therefore, I will assume that any noncausal sentence of $\mathscr{L}$ is derivable from a set of identities (which may be analytic or synthetic). This assumption will somewhat simplify the conditions and discussions presented below. It implies that, if L_1 n-eq L_2, then $L_1 \leftrightarrow L_2$ is derivable from a set of true identities of $\mathscr{L}$.

The assumption just made slightly restricts the class of theories to be considered. A more general class of theories would be obtained by allowing the possibility of noncausal sentences which are not derivable from synthetic identities plus analytic sentences. However, in practice, most, if not all, dynamic theories will satisfy the more specialized assumption made above. At any rate, this assumption is not pernicious. At the end of this section I will indicate how all of the work of this section can be generalized in such a way as to avoid this simplifying assumption.

Since $\mathbf{T}$ is a dynamic theory, it has the form $\mathbf{T} = \mathbf{F} \cup \mathbf{I} \cup \mathbf{D}$, where $\mathbf{F}$ is the set of fundamental law-sentences, $\mathbf{I}$ is the set of identities, and $\mathbf{D}$ is the set of derivative law-sentences. Of course, this form implies that any law-sentence of $\mathbf{T}$ is either in $\mathbf{F}$ or in $\mathbf{D}$. As was indicated in Chapter 2, not all deductive-nomological derivations are explanatory, i.e., represent causal explanations. However, suitable derivations from the fundamental law-sentences of a good theory presumably are explanatory. In general, these derivations will use an explanans which may contain fundamental law-sentences, identities, and analytic sentences. As always, the analytic sentences in the explanans can be removed to produce a new derivation of the same explanandum from fundamental law-sentences and identities. In order to characterize unified theories, I must now impose additional restrictions on the form of explanatory derivations and on the languages of the theories.

Let S be a sentence of $\mathscr{L}$, i.e., S is formulated in terms of the predicates in $\mathscr{L}$ plus the logical symbols of the language of $\mathbf{T}$. Let Γ be a set of sentences of $\mathscr{L}$. A derivation of S from Γ is a *minimal derivation* if and only if S is not derivable from any proper subset of Γ. Now let Γ be a subset of $\mathbf{F} \cup \mathbf{I}$, and let L be a law-sentence in $\mathbf{T}$. Then, a derivation of L from Γ is an *explanatory derivation* if and only if this derivation is D-N and minimal, and no law-sentence of $\mathscr{L}$ n-equivalent to any member of Γ is derivable from L. When there exists in $\mathbf{T}$ an explanatory derivation of L, then L is *explainable in* $\mathbf{T}$.

An explanatory derivation is a special kind of D-N derivation within a theory $\mathbf{T}$. Since it is a minimal derivation, its explanans contains no superfluous law-sentences or identities. Moreover, the law-sentences in its explanans are all fundamental law-sentences of $\mathbf{T}$. As will become clear below, it is intended that the explanandum L be a derivative law-sentence of **T**. Now it would be contrary to the meanings of the terms to have a fundamental law explainable by a derivative law of a theory. Therefore, it is natural to require that no law-sentence n-eq to any member of Γ be derivable from L. Yet, this definition of explanatory derivation is still quite liberal. I certainly do not claim that it provides a complete characterization of the notion of causal explanation. I do believe that it will adequately serve our purposes when used within the context of well-constructed, unified theories.

It is now necessary to impose some additional requirements on the language and the domain of **T**. A predicate which is a single predicate term is a *simple predicate*, and any predicate not logically equivalent to a simple predicate is a *complex predicate*. Let us now suppose that there is a distinguished subset of simple predicates of $\mathscr{L}$ which are called *basic*, and which satisfy the condition that there is no true identity between any basic predicate and any complex predicate definable only in terms of other basic predicates (plus logical terms). There may be a true identity between two different basic predicates.

Any kind or attribute which is denoted by a basic predicate is a *basic kind* or a *basic attribute*. Any other kinds or attributes of **T** are *derivative*, and it is required that the derivative kinds and attributes be denoted by complex predicates defined in terms of basic predicates of $\mathscr{L}$. These complex predicates are *derivative predicates*. A complex, derivative predicate may denote a kind or attribute which is also named by a simple predicate of $\mathscr{L}$. If this occurs, the simple predicate is also a derivative predicate.

These conditions may seem complicated, but their import is rather simple and natural. Any kind or attribute of **T** is either basic or derivative. Any basic kind or attribute is denoted by a basic predicate, which is a simple predicate. No basic kind or attribute is denoted by a complex predicate definable only in terms of other basic predicates. However, a basic kind or attribute may be denoted by more than one basic predicate. Thus, basic predicates are not necessarily logically primitive in the conventional sense of 'logically primitive'. The above conditions also require that any derivative kind or attribute must be denoted by a complex predicate defined in terms of basic predicates. Thus, no derivative kind or attribute is denoted only by a simple predicate of $\mathscr{L}$. Any predicate, whether simple or complex, which denotes a derivative kind or attribute, is a derivative predicate.

In general the form of the definitions of the complex derivative predicates will depend on the nature of the theory in which they are used. However, it is assumed that a thing-predicate is not definable only in terms of attribute-predicates, nor an attribute-predicate only in terms of thing-predicates. This restriction on allowable definitions is reasonable in the light of my discussion of natural classification systems in Chapter 3, Section C. In fact, I believe that all of the above restrictions are reasonable requirements to impose on the structure of a well-systematized dynamic theory. It is easy to see that these conditions will usually be satisfied by the kinds of theories discussed in Chapter 3.

Now let α be a simple predicate of $\mathscr{L}$, and let ϕ be a simple predicate of $\mathscr{L}$ or a complex predicate of $\mathscr{L}$ defined in terms of simple predicates other than α. Suppose that

$$(x_1) \cdots (x_n)(\alpha x_1 \cdots x_n \leftrightarrow \phi x_1 \cdots x_n)$$

is a law-sentence of **T**. Then α is said to occur as an *isolated correlate* in this law-sentence.

Finally, we say that a predicate α occurs *essentially* in a sentence S if and only if there is no sentence logically equivalent to S which has no occurrence of α. In the following, the word 'occurrence' will mean essential occurrence unless otherwise specified.

Using all of this terminology, I can now state several conditions which should be satisfied by any unified dynamic theory; indeed, several of these conditions should be satisfied by any good theory using the kind of language I have described.

U_1. Each simple predicate in $\mathscr{L}$ occurs in some law-sentence of **T**.

U_2. The extension of any predicate of $\mathscr{L}$ is (at any time) a subset of *Dom*, or a set of ordered n-tuples of elements of *Dom*.

U_3. **T** is closed under derivation of law-sentences of $\mathscr{L}$.

U_4. Each law-sentence in **D** is explainable in **T**.

U_5. No law-sentence in **F** is explainable in **T**.

U_6. Any law-sentence of $\mathscr{L}$ which is n-equivalent to a member of **F** is also a member of **F**.

U_7. Each law-sentence L_1 in **F** has an occurrence of a simple predicate of $\mathscr{L}$ which also occurs in another law-sentence L_2 of **F** such that L_1 is not n-equivalent to L_2.

U_8. No basic predicate of $\mathscr{L}$ occurs in law-sentences of **F** only as an isolated correlate.

U_9. All true identities between predicates of $\mathscr{L}$ are included in **I**.

U_{10}. If α is an n-place simple attribute-predicate of $\mathscr{L}$ such that the k-th component of any ordered n-tuple satisfying α is always of a derivative kind, then α is a derivative predicate (and hence names a derivative attribute, which is also denoted by some complex attribute-predicate defined in terms of basic predicates).

2. I will now attempt to motivate the conditions U_1–U_{10}. First of all, U_1 merely requires that $\mathscr{L}$ contain no simple predicates which are not actually used in the formulation of the law-sentences of **T**. This condition is stated for the sake of formal elegance; it has no effect on the substantive import of the theory. Since it can do no harm, it might as well be included among the necessary conditions for unity.

U_2 is an important condition which prevents the terms of $\mathscr{L}$ from making reference to any extraneous things or attributes beyond the scope of **T**. A theory which did not satisfy this condition could not be said to be unified. For example, in Chapter 3, Section B, I described some general characteristics of a theory of chemistry. This description was in terms of the idealized notion of empirically smallest samples of substances. However, roughly speaking, we can say that classical chemistry is concerned with the various kinds of substances and with the properties and behavior of these substances. But we might find a chemistry reference book or text which says that sodium cyanide is poisonous to rats. *Poisonous to rats* is a relational attribute which refers to *rats*. Rats may be included in the domain of some biological theory, but they are not a chemical substance and hence are not included in the domain of a purely chemical theory. The kinds of things in the domain of a unified chemical theory are the various kinds of chemical substances. Reference to rats could not play an integral role in the context of such a theory.

Actually, U_2 is a rather weak condition. In the above example we could force the satisfaction of U_2 by expanding the domain of the chemical theory to include rats. This procedure would seem *ad hoc*, and it is not likely to be done. Moreover, depending on various aspects of the theory, it might result in violations of some of the other conditions for unified theories. However, I cannot absolutely rule out a procedure of this kind. It is for this, and other reasons, that the conditions are claimed only to be necessary conditions for unified theories. U_2 is nonetheless a very important condition, and it will play an essential role in some later discussions.

U_3 states that if L is a law-sentence of $\mathscr{L}$, and if L is derivable from members of **T**, then L is also a member of **T**. One might be tempted to require that *all* law-sentences formulated in $\mathscr{L}$ be included in **T**. To require this would be to require that **T** be empirically complete in at least one sense. However, I do not see any reason why this empirical completeness requirement should be a requirement for unity. On the other hand, if L is also derivable from members of **T**, then one could hardly consider **T** to be unified if it did not also contain L. This is the intuitive reason for U_3. In addition, U_3 plays an important role in some further considerations below.

U_4 and U_5 are not really new conditions; they were imposed on all dynamic theories in Chapter 2, Section C. U_4 merely states that each derivative law-sentence of **T** is in fact explainable in **T**. U_5 states that no fundamental law-sentence of **T** is explainable in **T**. This is a relative sense of 'fundamental' because some or all of the fundamental law-sentences of **T** may be explainable by the fundamental law-sentences of some other theory. Since a law-sentence is a causal sentence, it is always in principle subject to a causal explanation.

It should also be noted that U_5 does *not* imply that the sentences in **F** $\cup$ **I** are logically independent of each other. U_5 says that no member of **F** is explainable in **T**, but this does not necessarily imply that no member of **F** is derivable from other members of **F** $\cup$ **I**. The following two lemmas show the relationships between the premises and conclusions of such derivations.

LEMMA 1. Let F, $F' \in$ **F**, and suppose that F' is derivable from F. Then F n-eq F'.

Proof. Let $F \vdash F'$ denote the derivation of F' from F. This must be a minimal derivation, for otherwise F' would be an analytic sentence and hence not a law-sentence. But $F \vdash F'$ is, by U_5, not an explanatory derivation. Hence, there is some law-sentence F'' of $\mathscr{L}$ such that F'' n-eq F and such that $F' \vdash F''$. But we have assumed that $\mathscr{L}$ is such that all n-equivalences are derivable from true identities of $\mathscr{L}$. Hence there is a set Δ of true identities of $\mathscr{L}$ such that $\Delta \vdash F \leftrightarrow F''$. By the Deduction Theorem of logic, from $F' \vdash F''$, we obtain $\vdash F' \rightarrow F''$. Hence, $\Delta \vdash F' \rightarrow F$. Also by the Deduction Theorem and $F \vdash F'$, we obtain $\vdash F \rightarrow F'$. Hence, $\Delta \vdash F \leftrightarrow F'$, so F n-eq F'.

LEMMA 2. Let $\Gamma \subseteq$ **F**, Γ nonempty, $\Delta \subseteq$ **I**, $F \in$ **F**, and suppose that $\Gamma \cup \Delta \vdash F$ is a minimal derivation. Then F is n-equivalent to a member of Γ.

Proof. Since $F \in$ **F**, by U_5 $\Gamma \cup \Delta \vdash F$ is not an explanatory derivation. Since this is a minimal derivation, there is some F' of $\mathscr{L}$ such that F' is n-equivalent to some member of Γ and such that $F \vdash F'$. By U_6 it follows that $F' \in$ **F**. Hence, both F, $F' \in$ **F**, and $F \vdash F'$. By *Lemma 1*, F n-eq F', so F is also n-equivalent to some member of Γ.

It is customary when axiomatizing a theory to try to obtain a set of axioms which are logically independent of each other. It is also customary to consider the fundamental law-sentences of a theory to be somewhat analogous to axioms in the sense that they are not explainable in that theory. Unfortunately, not all derivations represent explanations. Moreover, different law-sentences can state the same law. Thus, we cannot make the tempting assumption that the members of **F** are logically independent of each other. On the other hand, we should require that they be explanatorily independent of each other. U_5 accomplishes this, but it leaves us in the dark about the nature of derivations of fundamental law-sentences from other fundamental law-sentences and identities. Intuitively, one would expect the conclusions of such derivations to be intimately related to the premises, for otherwise the derivations might appear to represent causal explanations. Lemmas 1 and 2 exhibit the intimate relations one intuitively expects. In effect, these lemmas show that such derivations do not produce any new laws, for their conclusions are law-sentences which are n-equivalent to some law-sentence in the premises. It is interesting to note that I have made use of U_6 in the process of proving Lemma 2.

The motivation for U_6 should by now be fairly clear. Suppose that $F \in$ **F** and that F' n-eq F, where F' is any law-sentence of $\mathscr{L}$. Then there is a set Δ of true identities from which $F' \leftrightarrow F$ is derivable. By U_9, Δ is a subset of **I**, and $\Delta \cup \{F\} \vdash F'$. Then, by U_3, F' is a law-sentence of **T**. Thus, F' must be either in **F** or in **D**. But F' states the same law as F, so we are forced to require that F' be in **F**. We certainly cannot allow one statement of a law to be a fundamental law-sentence while a different statement of the same law is a derivative law-sentence.

Of course, this defense of U_6 makes use of U_9, which requires that all true identities of $\mathscr{L}$ be included in **T**. In discussing U_3 I maintained that it is not necessary for unity to have all law-sentences of $\mathscr{L}$ in **T**. But the situation is different in the case of the identities. U_9 does not require completeness in the sense that all laws statable in $\mathscr{L}$ are included in **T**. If we add an identity to a theory, we may thereby make possible the derivation of new law-sentences. But the set of fundamental laws, and hence the set of fundamental causal connections stated by the theory, will remain the same. The aim of including all true identities of $\mathscr{L}$ in $\mathbf{I}$ is to guarantee that the theory will indicate the equivalence of two different descriptions of the same phenomena. If this aim is not realized, then the theory's ontology (see Chapter 2, Section C) will appear more diverse than it actually is, and thus the theory will not provide a truly unified description of the kinds and attributes to which it refers.

If U_9 is not satisfied, it is very likely that $\mathbf{T}$ will contain law-sentences which are really n-equivalent, but such that their n-equivalence is not demonstrable within $\mathbf{T}$. The analogous circumstance is likely to arise for certain e-equivalent derivations. Moreover, if a true identity is not stated, then (especially in the case of attributes) it is likely to be misconstrued as a nomological co-extensionality and treated as a law-sentence of $\mathbf{T}$. Then $\mathbf{T}$ will contain a pseudo-law, and fruitless efforts might be expended in a vain attempt to explain this pseudo-law. In view of all of these considerations, I conclude that U_9 is necessary in order for $\mathbf{T}$ to be unified. Thus, for example, if $\mathbf{T}$ is a theory whose language refers both to gravitational mass and to inertial mass, and if these are identical, then $\mathbf{T}$ should say that they are by containing the appropriate identity in $\mathbf{I}$. Otherwise, $\mathbf{T}$ would not provide a unified account of the phenomena with which it is concerned.

I believe that the motivations for U_7 and U_8 should be fairly obvious. U_7 requires that the fundamental laws of $\mathbf{T}$ be linked together in a certain sense. This condition prevents $\mathbf{T}$ from containing an isolated law-sentence which can play no part in explanatory derivations in $\mathbf{T}$. A somewhat similar condition occurs in Bunge (1967, pp. 394–395), although Bunge's condition is simpler to state since it does not take account of the possibility of n-equivalent law-sentences in $\mathbf{T}$.

U_8 prevents $\mathscr{L}$ from referring to a basic kind or a basic attribute which, from the point of view of $\mathbf{T}$, is superfluous. In Chapter 5 I discussed the difficulty which is sometimes encountered in practice in deciding whether a given co-extensionality is an identity or a correlation law. U_8 is a kind of limited Occam's razor principle which settles this difficulty in certain special cases. Suppose that α is a basic predicate which occurs in law-sentences of $\mathbf{F}$ only as an isolated correlate. For instance, suppose that α is so correlated with ϕ. Then, by the definition of 'isolated correlate', α and ϕ are nomologically co-extensional, and thus not identical. But if α and ϕ are not identical, then one would expect that α would play some role in **T** essentially different from a role which could equally well be played by ϕ. Thus, one would expect α to occur in some member of **F** other than as an isolated correlate. Since, by hypothesis, α does not do this, there is no need to have α in $\mathscr{L}$. Neither α, nor any other basic predicate related to α by an identity, plays an independent, integral role in the theory. The predicate α is therefore excluded from $\mathscr{L}$. It should be noticed that α might occur as an isolated correlate with ϕ and with ψ, etc. If this occurs, α should still be excluded, although in this case we might need to include in **F** the appropriate correlation law-sentences between ϕ, ψ, etc.

It might be objected that isolated correlates should be allowed in **F** because an α and a ϕ might in fact be nomologically co-extensional. However, if this happened, then the correlation law-sentence should in principle be explainable by the reduction of **T** to some other theory, or by some revised and perhaps strengthened version of **T** itself. In either case it seems almost certain that **T** would be modified in such a way that α would then occur in some fundamental law-sentence other than as an isolated correlate. At any rate, it does not appear that an isolated correlate has a place in a unified theory. Although I believe that U_8 is necessary for unity, it should be mentioned that this condition is not required for the argument of the next section.

Since U_9 has already been discussed, I must now motivate U_{10}, which is perhaps the least obvious of all of the conditions. To see the rationale for U_{10}, recall the distinction between basic and derivative kinds and attributes. Within **T**, the basic kinds are those kinds of elements of *Dom* which are denoted by simple predicates and cannot be denoted by complex predicates defined only in terms of other basic predicates. In general, *Dom* will also contain derivative kinds which must be denotable by complex predicates definable in terms of basic predicates. Almost all dynamic theories, and especially unified dynamic theories, will possess this distinction because it is desirable that the ontology of the theory be constructed from a minimal number of independent, basic kinds. When the theory's ontology is constructed in this way, we obtain a more unified picture of the relationships between the various kinds of things in the domain. However, in order for such a construction to accomplish its aim, the attribute-predicates of $\mathscr{L}$ must also be involved in the construction in an appropriate manner.

The attributes referred to in $\mathscr{L}$ are also part of the theory's ontology. These attributes are distinguished into basic and derivative attributes in the same way as the kinds are so distinguished. The aim is to obtain a unified construction of the entire ontology of T in terms of the basic kinds and basic attributes. In addition, these independent basic kinds and attributes should be interrelated into a unified system. If a property or quantity is never possessed by any basic kind, or if a component of a relation never has a relatum which is a basic kind, then this property, quantity, or relation plays no role in the basic ontology of the theory. An attribute with this nature should therefore be a derivative attribute, and U_{10} requires it to be one. The result should now be quite obvious. Although a basic attribute *may* be possessed by a derivative kind, it *must* be possessed by at least one basic kind.

It is important to recognize that T is not required to have any derivative kinds or attributes, although most theories do. Any derivative kinds or attributes which do occur in T are supposed to play a secondary role. One would expect that all fundamental law-sentences of T should be law-sentences about basic kinds and about attributes which are possessed by basic kinds. U_{10} guarantees that this is the case. All together, U_1–U_{10} go a long way towards requiring that T has unified and interlocking systems of laws, ontology, and intratheoretic explanations. I believe that most theoreticians would agree with me that U_1–U_{10} must be satisfied in order for a dynamic theory to be unified. These conditions are therefore *necessary* conditions for a theory to be unified; it is not claimed that they are *sufficient* for unity. I doubt that any set of general logical and semantical conditions would be sufficient for unity. Whether or not a particular theory is unified will in part depend on particular substantive features of that theory. Fortunately, the necessary conditions U_1–U_{10} have important implications concerning the unification of science.

Before concluding this section, it is necessary to reconsider the role of noncausal sentences in the preceding discussions. Early in this section I made the simplifying assumption that all noncausal sentences of T are derivable from identities plus analytic sentences. This simplifying assumption can be avoided by making appropriate revisions. First one allows T to have the form $F \cup N \cup D$, where N is now the set of non-causal sentences of the theory. Explanatory derivations may have explanans which are subsets of $F \cup N$, and the definition of 'explanatory derivation' is revised accordingly. Also U_9 is revised by replacing 'I' by 'N'. Of course, this revised version of U_9 includes the original version as an important special case. The justification of the revised U_9 is essentially the same as that given for the original version. These changes do not affect most of the arguments of this section. The changes are relevant to a few of the arguments. In these cases it is easy to see how to revise the arguments slightly in order to accommodate the changes involving 'I' and 'N'.

C. UNIFICATION BY MICROREDUCTION

1. In this section I argue that, if T_2 can be microreduced to T_1, then the only satisfactory way to unify these two theories is by a microreduction to T_1, or to a theory which is a modified formulation of T_1. This argument provides the foundation for a program for the unification of science based on successive microreductions of various theories. The first section of this chapter contains a brief review of the Oppenheim-Putnam program for unification by microreductions. I mentioned that several questions can be raised about such a program, in particular: (i) What conditions must be satisfied in order for a theory to be unified? (ii) What conditions must be satisfied in order to achieve an adequate microreduction? (iii) Why should the unification program be based on successive microreductions? (iv) What reasons are there for believing that the proper empirical relationships will exist between the various branches of science?

As stated in Section A, Oppenheim and Putnam reviewed scientific developments in order to provide some answers to (iv). In the next chapter I will provide some further considerations relevant to (iv). Also I have provided an answer to (ii) in earlier chapters of this book. My conditions for adequate microreductions are considerably different, and far more detailed than those presented by Oppenheim and Putnam. No attempt is made by Oppenheim and Putnam to answer question (i). Fortunately, we now have at least a partial answer to this question in the form of U_1–U_{10} of the previous section. It is therefore now possible to give an answer to (iii).

2. In general, if two theories, T_1, T_2, are to be unified, it is necessary to construct some *unified* theory T which at least contains all of the laws and identities of T_1 and T_2. This goal clearly cannot be accomplished with all conceivable pairs of theories. Furthermore, even when two theories can be unified, it is not necessarily the case that they can be unified by a microreduction. The argument of this section is therefore limited to pairs of theories, T_1, T_2, such that T_2 can be microreduced to T_1. I also assume that T_1 and T_2 are *unified theories*. This is a reasonable

and realistic assumption in view of the conditions for adequate micro-reductions.

Now it is possible that T_1 and T_2 could both be reduced to some third theory T_3, which is a unified theory. The result of these reductions might be a unification of all three theories, but it would not be simply a unification of T_1 and T_2. Therefore, I will suppose that the set of kinds and attributes of T is the union of the set of kinds and attributes of T_1 with the set of kinds and attributes of T_2.

$T = F \cup I \cup D$ has a domain, *Dom*, and a set of nonlogical predicates $\mathscr{L}$. We have just seen that $Dom = Dom_1 \cup Dom_2$. T must contain all laws and identities of T_1 and of T_2. In addition, because of U_9, I will at least contain certain thing-identities stating that certain kinds of things in Dom_2 are the same as certain kinds of things in Dom_1. In order to guarantee that all of the appropriate laws and identities can be stated in T, I will assume that $\mathscr{L}_1 \cup \mathscr{L}_2 \subseteq \mathscr{L}$. This is a technical simplification which produces no essential loss of generality. Note that this condition does not specify the basic predicates of $\mathscr{L}$.

T_2 can be microreduced to T_1, and T_1 is a theory with structured wholes. Also T_1 is unified, so the basic kinds in Dom_1 are the various kinds of possible parts, and the derivative kinds in Dom_1 are the various kinds of structured wholes (composed of two or more parts). The detailed characteristics of theories with structured wholes were described in Chapter 3. T_1 possesses these characteristics, as well as those of unified theories. Among other things, T_1 contains law-sentences which state the conditions under which the various kinds of structures do or do not form.

In contrast to T_1, we suppose that T_2 is a theory which has only basic kinds and basic attributes. The fact that T_2 can be microreduced to T_1 implies that each kind in Dom_2 is identical with some kind in Dom_1. In order for the argument to proceed more smoothly, I will assume that the microreduction of T_2 to T_1 satisfies a more specialized and familiar condition. It is assumed that each kind in Dom_2 is identical with some derivative kind in Dom_1, and conversely. This condition is frequently taken as a defining characteristic of microreductions. In earlier chapters I did not consider it as such because I wished to have a slightly more general set of requirements for microreductions. However, in the present context we are primarily concerned with the relative merits of microreductions versus macroreductions. It is therefore appropriate to assume as an hypothesis of the argument that the set of kinds of T_2 is identical with the set of derivative kinds of T_1.

3. The problem situation which concerns us is now easily summarized. We have two unified theories, T_1 and T_2. T_1 is a theory with structured wholes. The basic kinds in Dom_1 are the possible parts, and the derivative kinds are the various kinds of structured wholes. T_2 is a theory which has only basic kinds and basic attributes. Each derivative kind in Dom_1 is a basic kind in Dom_2, and conversely. T_2 is microreducible to T_1.

We wish to construct a unified theory T with a set of kinds and attributes which is the union of the kinds and attributes of T_1 with those of T_2. The set $\mathscr{L}$ of nonlogical predicates of T contains all of the nonlogical predicates in $\mathscr{L}_1 \cup \mathscr{L}_2$. T contains all of the laws and identities which are in either T_1 or in T_2.

Given this problem situation, many would say that the natural way to try to construct T is by means of the microreduction of T_2 to T_1. However, it is not immediately obvious that this microreduction does produce a unified theory. Moreover, there may be other ways of producing a unified T; indeed, the arguments of Schlesinger suggest that this may be the case. Finally, suppose that there are several ways of constructing a unified T, among which one is by a microreduction. Then it is by no means obvious that the microreduction would provide the best way. In fact, Schlesinger claims that the common preference for microreductive unifications is unjustifiable. Probably his view is partly the result of the fact that he does not consider detailed conditions for the unity of a theory. At any rate, I accept his view as a challenge to justify the use of microreductions as a unifying procedure. Given the problem situation just described, I will consider all possible ways to construct a T which unifies T_1 and T_2. The possibilities are divided into three general cases. I will argue that the only satisfactory way is given by the third case, which corresponds to performing a microreduction.

Case 1. Let T be a theory whose basic kinds are a subset of those of T_2. Constructing this kind of theory would, at least in most cases, amount to performing what Schlesinger considers to be a macroreduction of T_1 to T_2. Since the basic kinds in Dom are a subset of those of T_2, the derivative kinds in Dom include all of the basic kinds of Dom_1. These derivative kinds in Dom must be denotable by complex predicates definable in terms of basic predicates of $\mathscr{L}$. Likewise, any derivative attributes of T must be denotable by definitions in terms of the basic predicates of $\mathscr{L}$. Except for extremely trivial theories, it is unlikely that definitions of the derivative predicates of $\mathscr{L}$ can be obtained.

Consider a basic thing-predicate 'B' in $\mathscr{L}_1$. The kind denoted by

'B' must now be denoted by a derivative thing-predicate of $\mathscr{L}$. Recall that $\mathscr{L}$ uses no definite descriptions, and that a thing-predicate is not definable only in terms of attribute-predicates. Thus, 'B' must denote the same kind as some predicate $\phi(W_1, ..., W_m, A_1, ..., A_n)$, where '$W_1$', ..., '$W_m$', are basic thing-predicates of $\mathscr{L}$, and 'A_1', ..., 'A_n', are basic attribute-predicates of $\mathscr{L}$. In Chapter 3 I showed in detail how structures can be so defined in terms of their parts plus structural relation-predicates. Within $\mathscr{L}_1$ there are basic predicates for the various kinds of parts, and there are also structural relation-predicates. It is then possible to define complex predicates denoting the structures. The procedures used will be those described in Chapter 3. However, these procedures cannot be used in $\mathscr{L}$; instead $\mathscr{L}$ must somehow accomplish the reverse procedure, and it is very hard to see how this could be done if Dom_1 has very many kinds of parts and structures. Many, if not all, structural relation-predicates of $\mathscr{L}_1$ denote relations which apply only to the basic elements of Dom_1, i.e., the kinds of parts in this domain. In Dom these parts are derivative kinds, so by U_{10}, most, if not all, structural relation-predicates would have to be derivative predicates of $\mathscr{L}$. But the basic predicates of $\mathscr{L}$ refer to structures and to attributes possessed by structures. It is thus very unlikely that the structural relation-predicates could be defined in terms of these basic predicates of $\mathscr{L}$. Also the part-relation, 'Pt', of Chapter 5, Section E, probably could not be defined in terms of basic predicates of $\mathscr{L}$. For these reasons, it is very unlikely that there could be a suitable predicate $\phi(W_1, ..., W_m, A_1, ..., A_n)$ which denotes the same kind of thing as 'B'.

I realize that the argument just presented is not logically rigorous. Nevertheless, this argument at least shows that, within $\mathscr{L}$, it would be difficult, if not impossible, to obtain adequate definitions of parts in terms of predicates denoting wholes and attributes possessed by wholes. Furthermore, if the definitions are obtained, then a law-sentence about uncombined parts must be n-equivalent to a law-sentence about structures composed of parts. When a microreduction is accomplished, it has the result that law-sentences about wholes are n-equivalent to law-sentences about parts under the external boundary conditions plus the boundary conditions imposed by the relevant structural arrangements. This is not surprising because the wholes are composed of the parts. In $\mathbf{T}$, on the other hand, a law-sentence about uncombined parts must state the same law as a law-sentence about wholes. This result hardly seems possible because a particular kind of basic element in Dom_1 can usually be a part in many different kinds of structured wholes under different conditions. Whenever a basic element of Dom_1 is combined into a structured whole, it is under special boundary conditions imposed by the relevant structural conditions. It is hard to imagine how a law-sentence about the behavior of an uncombined kind of part could state the same law as law-sentences about the behavior of various structured wholes in which this kind of part is under various types of special boundary conditions. These remarks do not provide a rigorous impossibility proof, but they do provide further reason to doubt the possibility of the kind of unified theory we are now considering. They also provide further reason to doubt that the basic kinds in Dom_1 can be denoted by definitions in terms of the basic predicates of $\mathscr{L}$.

Let us now assume that the difficulties described above can be overcome. T must contain all of the laws and identities of both T_1 and T_2. In particular, the fundamental laws of T_1 must be formulated somehow in T. If all of these formulations are fundamental law-sentences of T, then they will be sufficient, together with the identities of T, to derive all other law-sentences of T. This is because T_2 is microreducible to T_1, and because of U_9 plus the fact that I contains any identity which is in I_1 or in I_2. But in this situation the fundamental law-sentences of T will simply be $\mathscr{L}$-formulations of those in F_1. Basically T will provide a microreduction of T_2 to T_1, although this microreduction will be formulated in a counter-intuitive, and unnecessarily complicated way, using basic thing-predicates denoting basic kinds in Dom_2. In effect, T is in this case not a macroreduction, but rather an awkward microreduction, which might as well be replaced by a normal microreduction.

The bizarre microreduction just described may be amusing, but it is hardly of theoretical or practical significance. So let us now assume instead that T has some derivative law-sentences which are $\mathscr{L}$-formulations of F_1 law-sentences. For instance, consider an F_1 law-sentence which states that a certain basic kind in Dom_1 has a certain property, and suppose that its $\mathscr{L}$-formulation ψ is a derivative law-sentence of T. Then ψ must be derivable in T from a set of premises containing fundamental law-sentences of T referring to structured wholes and the attributes and behavior of these wholes. Such a derivation should represent a causal explanation in T of ψ. But this would mean that, in some sense, a property of an uncombined thing is caused by attributes of possible structures of which this thing may or may not be a possible part. Within the context of a good dynamic theory, a causal explanation of why a certain kind of thing has a certain property should be an explanation which somehow shows how this property is caused by other attributes of this kind of thing. Our supposed derivation of ψ does not accomplish this, and it seems to me that this derivation cannot represent a causal explanation. If a properly constructed derivation of a derivative law-sentence of a theory from fundamental law-sentences and identities of that theory is not explanatory, then the theory is presumably faulty in some respect. I think that we must conclude that the type of theory now being considered is not acceptable, and hence does not provide an acceptable unification of $\mathbf{T}_1$ and $\mathbf{T}_2$.

Let us consider another reason which supports the view that $\mathbf{T}$ is unsatisfactory. Since $\mathbf{T}_1$ is a theory with structured wholes, it has law-sentences stating the conditions under which various structures do or do not form. In $\mathbf{T}$ it is likely that their $\mathscr{L}$-formulations will be derivative law-sentences. They will then be derived from law-sentences about non-structural attributes of structured wholes. Such attributes of a structure may help to determine its future persistence under certain conditions; but (except for perhaps limiting cases) they cannot be causal conditions for the formation of structures. Therefore, within $\mathbf{T}$ the derivations of the law-sentences of formation of structures would most likely fail to be explanatory.

I have now made several observations about the difficulty of constructing $\mathbf{T}$, and about various undesirable features it would have. Perhaps none of these observations alone is sufficient to rule out the possibility of a unifying theory of the type considered in *Case 1*. However, when all of these observations are considered together, I believe that they show that the $\mathbf{T}$ of *Case 1* is probably impossible to construct (with nontrivial $\mathbf{T}_1$ and $\mathbf{T}_2$), and that, if it is constructable, then it would be an unacceptable theory.

Case 2. Suppose that $\mathbf{T}$ is a theory whose basic kinds consist of a nonempty subset of the basic kinds in Dom_1 together with a nonempty subset of the basic kinds in Dom_2, i.e., some of the basic kinds in Dom are kinds of parts and some are kinds of structured wholes. Let 'W' be a basic predicate of $\mathscr{L}$ which denotes a kind of structured whole. Then no complex predicate of $\mathscr{L}$ defined only in terms of other basic predicates of $\mathscr{L}$ can have the same denotation as 'W'. But any kind or attribute referred to by an $\mathscr{L}$-predicate is a kind or attribute referred to by an $\mathscr{L}_1$-predicate or by an $\mathscr{L}_2$-predicate. $\mathscr{L}_1$ has structural relation-predicates which are used in the definitions of various kinds of possible structured wholes. As we saw above, these structural relation-predicates are likely to be derivative in $\mathscr{L}$, and if they are, then they must be definable in terms of basic predicates of $\mathscr{L}$. The kind of structured whole denoted by 'W' will thus also be denoted by a complex predicate defined in terms of structural relation-predicates plus predicates denoting the various kinds of parts of W. If any structural relation-predicates are derivative, their definitions can be used in this complex predicate. Hence, the kind of whole W will also be denoted in $\mathscr{L}$ by a predicate χ, such that χ is defined in terms of basic attribute-predicates of $\mathscr{L}$ plus predicates denoting the parts of W. But the predicates denoting these kinds of parts cannot all be basic predicates of $\mathscr{L}$, for otherwise 'W' would not be basic. Hence, at least one kind of part of W must be a derivative kind of Dom, and it will be denoted by a complex derivative predicate ρ in $\mathscr{L}$. If ρ is not used in χ, it can be substituted into χ. Since ρ is a thing-predicate, it cannot be defined only in terms of attribute-predicates. Thus, the definition of ρ must use some basic thing-predicates of $\mathscr{L}$.

Now suppose first that ρ is the only derivative thing-predicate used in χ. Then, if the basic thing-predicates used in ρ are all different from 'W', then again 'W' would not be basic, because it would be definable in terms of other basic predicates of $\mathscr{L}$. Therefore, the complex thing-predicate ρ must use 'W' in its definition. On the other hand, if ρ is not the only derivative thing-predicate used in χ, then there will be at least one derivative thing-predicate used in χ for which we obtain the same conclusion by this argument. Thus, $\mathscr{L}$ will contain at least one derivative thing-predicate, denoting a basic kind of Dom_1, whose definition uses a basic thing-predicate of $\mathscr{L}$ which denotes a basic kind of Dom_2. In other words, at least one kind of part will be defined in $\mathscr{L}$ using a definition which refers to at least one kind of whole. Needless to say, $\mathscr{L}$ could have many derivative thing-predicates with this character.

The result just obtained implies that the **T** of *Case 2* will suffer from many of the same difficulties and peculiarities as those suffered by the **T** of *Case 1*. It will be difficult, if not impossible, to construct the definitions of the complex predicates denoting the various kinds of derivative parts in *Dom*. If the definitions are obtained, then some law-sentences about the behavior of uncombined derivative parts will be n-equivalent to law-sentences which are at least partly about the attributes and behavior of structured wholes. Also **T** will encounter problems with its explanations which are similar to the problems with explanations described previously in *Case 1*. In general, the severity of the problems encountered by **T** should be approximately proportional to the number of kinds of Dom_2 which are basic kinds of *Dom*. If most of the kinds of Dom_2 are basic kinds of *Dom*, then the **T** of *Case 2* will be very similar to the **T** of *Case 1*. If most of the kinds of Dom_2 are derivative kinds of *Dom*, then the **T** of *Case 2* will be very much like a microreduction. The more nearly **T** approximates a microreduction, the less likely it will be to encounter many severe problems. I conclude that any theory included in *Case 2*, if it could be constructed, would suffer from the same type of defects as those suffered by the theories of *Case 1*. A theory included in *Case 2* would be inferior to a microreduction.

Case 3. In this case we suppose that the basic kinds in *Dom* are a subset of the basic kinds in Dom_1. Correspondingly, we might as well assume that the basic thing-predicates of $\mathscr{L}$ are the basic thing-predicates of $\mathscr{L}_1$ denoting those basic kinds of Dom_1 which are now basic kinds of *Dom*. $\mathscr{L}$ must be capable of defining complex thing-predicates denoting the structures in Dom_1. Each of these structured wholes is identical with some basic kind in Dom_2. Because of U_9 **I** must contain identities between these complex thing-predicates and the corresponding basic thing-predicates of $\mathscr{L}_2$. Because of U_{10} no basic attribute-predicate of $\mathscr{L}$ can have any component which applies only to structured wholes. Thus, any basic attribute-predicate of $\mathscr{L}_2$ which applies (in any one of its components) only to basic kinds of Dom_2 must be denoted in $\mathscr{L}$ by a complex, derivative predicate. By U_9 **I** must contain identities between these complex attribute-predicates and the corresponding basic attribute-predicates of $\mathscr{L}_2$. In other words, any basic attribute-predicate of $\mathscr{L}$ must apply to basic kinds of *Dom*, i.e., the various kinds of possible parts of structured wholes.

The set of basic predicates of $\mathscr{L}$ has still not been completely characterized. This set of basic predicates may be exactly the same as that of $\mathscr{L}_1$, or it may be different. Since T_1 is assumed to be a well-formulated, unified theory with structured wholes, the latter possibility is not very likely. Moreover, in view of the above discussion, if the latter possibility did occur, the set of basic predicates of $\mathscr{L}$ would still be very similar to the corresponding set in $\mathscr{L}_1$. The basic thing-predicates of $\mathscr{L}$ must denote various kinds of possible parts, and the basic attribute-predicates of $\mathscr{L}$ must denote attributes possessed by these various kinds of parts. In any event, **T** must be a theory with structured wholes, and it must contain the law-sentences and identities of T_1. **T** must also contain the law-sentences of T_2, but we now see that they must be n-equivalent in $\mathscr{L}$ to law-sentences about structured wholes. The latter will, in turn, be n-equivalent to law sentences about parts and attributes of parts.

It has just been argued that the set of basic predicates of $\mathscr{L}$ must be the same as, or approximately the same as, the set of basic predicates of $\mathscr{L}_1$. In spite of this, it might be suggested that the fundamental law-sentences of T could be different from those of T_1. The explanatory derivations in T would then be different from those of T_1. This may happen to some extent. However, because of the identity, or near identity, between their sets of basic predicates, to whatever extent it is possible for F to differ from F_1, it should be possible to reformulate T_1 in the same manner. It would then also be possible to change the explanatory derivations in T_1 correspondingly. But since T_1 is a theory with structured wholes, its fundamental law-sentences and explanatory derivations must satisfy the rather rigid constraints imposed on such theories in Chapter 3. It therefore could not be modified very much, and even if it were, the result would be a bad theory with derivations which are not explanatory. Thus, F should also be very similar to, if not identical with, F_1.

The result of all of this discussion is that T should produce a unification by means of a microreduction to a theory which is the same as, or nearly the same as, T_1. This result is obtained from the characteristics of T_1 and T_2 and the relationship between their domains, together with some of the conditions for unified theories, plus the special hypothesis of *Case 3*. Under these conditions, if T_1 and T_2 are to be unified, then we have the result that this unification must be by means of a microreduction. Of course, these conditions do not guarantee that the microreduction is possible. However, I have also made the stronger assumption that T_2 is microreducible to T_1. In addition, it was assumed that T_1 and T_2 are unified theories. Therefore, if the microreduction of T_2 to T_1 is performed according to the procedures described in Chapter 5, and if all true identities of $\mathscr{L}$ are included in I, then we obtain a theory T which satisfies U_1–U_{10}.

Again I emphasize the obvious fact that these arguments are not logically rigorous. However, I believe that they are as rigorous as is necessary or feasible in this type of philosophical subject matter. These arguments show that, if T_2 is microreducible to T_1, then: (i) T_1 and T_2 can be unified by means of this microreduction, and (ii) any satisfactory unification of T_1 and T_2 must be by means of a microreduction to T_1, or to a theory which is an acceptable reformulation of T_1.

This conclusion provides a rational foundation for a program for the unification of science based on successive microreductions. If we have a sequence of theories which are successively microreducible to a lowest level theory, and if we wish to unify these theories, then this unification should be based on successive microreductions. Contrary to the opinion of Schlesinger, this kind of unification program is not the result of a mere prejudice in favor of microreductions; it has a rational foundation. In order to understand this rational foundation, one must consider the necessary conditions for a unified theory and carefully examine their implications.

Interestingly, it has also been claimed that a macroreduction of T_1 to T_2 may produce a theory just as simple as a microreduction of T_2 to T_1; see Schlesinger (1963, p. 48). The arguments given above under *Case 1* clearly contradict this claim. Unfortunately, Schlesinger does not consider any detailed conditions which are necessary for a unified theory, nor does he consider any detailed conditions for microreductions. His claim about the relative simplicity of microreductions and macroreductions is therefore not surprising.

One final remark is now in order. The reader may have initially wondered why the conditions for unified theories are stated in the elaborate way they are. Instead of using the distinctions between simple and complex predicates, and between basic and derivative predicates, it would have been much simpler to use the customary distinction between logically primitive and defined predicates. Unfortunately, the latter procedure results in some loss of generality, and limits the types of identities which can occur in the theory. This limitation would then make it very difficult, if not impossible, to discuss the various types of unified theories considered in *Cases 1, 2*, and *3*. For this, and other reasons, it is at least very convenient to characterize unified theories in the manner used in this chapter.

The use of microreductions for the purpose of theory unification has now been justified. We therefore have justification for giving serious consideration to a program for the unification of science based on successive microreductions. This program must overcome many obstacles – some empirical, some conceptual – if it is to be successful. In the next chapter I will discuss additional aspects of this program, and examine some of its most serious obstacles.

NOTE

[1] Massey (1973) contains interesting comments on the Carnap program and related issues. Massey's approach to these issues differs significantly from that taken in this book. Yet, among other things, he points out certain difficulties which seem to result from the use of set theoretical scientific languages.

CHAPTER 7

COMPLICATIONS AND OBSTACLES

This chapter examines various important kinds of complications and obstacles involved in a program for the unification of science by means of successive microreductions. The first section is concerned with complications resulting from the enormous variety of types of structures found in the world. Section B is concerned with special types of complications found in hierarchical structures. In the third section I critically examine some recent, subtle arguments of J. A. Fodor. Fodor's arguments are intended to undermine the kind of reductionistic unification program defended in this book. Although Fodor's arguments have some force, it is shown that they do not establish an insuperable obstacle for the unification program. Finally, in the last section I consider some special difficulties arising in the social sciences.

In recent years there have been many arguments by philosophers and scientists for or against the possibility of the reduction of certain types of theories to certain other types of theories. For instance, there has been much controversy concerning the reducibility of biology to the physical sciences, and concerning the reducibility of psychology to neurophysiology. In addition, there has been much discussion of the mind/body identity thesis, which has certain connections with the reducibility of psychology to neurophysiology. The arguments involved in these controversies often involve complex conceptual analysis and sometimes substantive features of certain scientific theories. It is neither desirable, nor necessary, nor feasible to pursue the many complexities of these specific controversies in this book. In this chapter I will discuss certain important, *general* complications and obstacles involved in the program for the unification of science through successive microreductions.

It should also be noted that this chapter is concerned with the unification of the dynamic theories of the various branches of science. Chapter 8 contains some discussion of the relationship between these dynamic theories and evolutionary and developmental theories.

A. VARIETY OF STRUCTURES AND THEORIES

In the previous chapter I presented some necessary conditions for the unity of a theory, and then I used these conditions to defend the use of microreductions as a theory-unifying procedure. It is therefore of some interest to consider a program for the unification of science by means of successive microreductions. It will be recalled from Chapter 6 that Oppenheim and Putnam had previously defended such a program as a working hypothesis. The Oppenheim-Putnam program utilized a simplified model of science in which the various branches of science are assumed to correspond to various levels of structural complexity. In their model the first level is that of elementary particles, above this is the level of atoms, then the level of molecules, the level of living cells, the level of multicellular organisms, and finally the level of social groups. Although this simple system of levels was suitable for their purpose, it is obviously not an accurate picture of the organization of current science.

If one looks at the directory of a large university, he will find an almost bewildering array of scientific departments and research groups. For example, in such a directory I find traditional departments of anthropology, astronomy, botany, chemistry, economics, microbiology, physics, etc. In addition, there are many specialized institutes and laboratories such as an algal physiology laboratory, an atmospheric science group, a biochemical institute, a fusion research center, a genetics foundation, a marine science institute, a nuclear studies center, a particle theory center, a plasma dynamics laboratory, a population research center, and many others. Clearly, modern scientific research is not organized according to some simple system of structural levels. Unification of science by successive microreductions might therefore appear to be hopeless because of the enormous variety of things in the world and the great number of theories about these kinds of things. Although the situation is indeed complicated, I do not believe that it is hopeless. By suitable elaborations of the idea of successive microreductions, I believe that the unification program can accommodate the large variety of structures and theories. In the remainder of this section I will discuss the kinds of elaborations which will probably be required.

Suppose that all of the kinds of things in the world could be classified into a linearly ordered sequence of domains, $Dom_1, Dom_2, \ldots, Dom_n$, governed by theories, $T_1, T_2, \ldots, T_n$, respectively. Suppose that, for each $i < n$, $Dom_i = Bas_i \cup Comp_i$, and that T_i is a theory with structured wholes. Also suppose that, for each $i > 1$, $Bas_i = Comp_{i-1}$. Then it would be natural to attempt a unification of these theories by means of uniform microreductions of the fundamental law-sentences of $\mathbf{T}_i$ to $\mathbf{T}_{i-1}$, for all $i > 1$. Fortunately, we already know that the world is far more interesting and complex than this simple, linear model suggests (Causey (1968a)). At the very least, the unification program must be able to handle a more complex array of domains and theories. Let us briefly consider some of the complexities of this array.

In the Oppenheim-Putnam model, the lowest domain consists of fundamental particles, and the next domain consists of atoms. The components of atoms, e.g., electrons, protons, and neutrons, were considered to be fundamental particles. At the present time, fundamental particle theory is still undergoing significant development. In particular, some particles which were previously considered fundamental are now believed by some theorists to be composed of other particles. To take one recent example, Feynman (1974) has argued that protons may be composed of the hypothesized "simpler" particles called "quarks". This is but one example of many theories which indicate that the relationship between atoms and "fundamental particles" is more complex than had previously been believed.

If we consider the kinds of things which are built up from atoms and molecules, the structural possibilities also become very extensive. Atoms not only exist in electrically neutral forms, such as Ar, but also in the forms of charged ions, such as Na^+. Structures composed of atoms (i.e., things on the molecular level), can also exist in ionized forms, e.g., SO_4^{--}. Under certain conditions it is also possible to form free radicals, such as the triphenylmethyl radical, $(C_6H_5)_3-C$. Radicals are usually not considered to be stable molecules, although they have a molecular-type structure in so far as they are composed of atoms chemically bonded together. These and other types of objects are now discussed in any good general chemistry textbook. Although they add complexities to the theories of atoms and of molecules, they do not essentially change the structural relationships between the domain of atoms and the domain of molecules. On the other hand, there are plasmas, composed largely of ions and fundamental particles. There are also thought to be neutron stars, with cores consisting of densely packed fundamental particles, as described in Levitt (1974, Chapter 7). Objects such as these require somewhat more extensive theoretical developments.

In addition, when we consider structures composed of molecules, we find a variety of general types. For instance, there are macroscopic samples of substances. In the solid state, these samples often, but not always, consist of molecules (or ions) arranged in the form of some particular type of crystalline structure. In the liquid and solid states, macroscopic samples of substances have their molecules arranged into other types of structures. Moreover, molecules can form various types of microscopic structures. For example, in water solution the ions of long-chain fatty acids form small structured aggregates called "micelles"; see Platt (1961, p. 355). Structures somewhat analogous to micelles are believed to be major components of cell membranes; see, for instance, Dowben (1969). Other types of microscopic complexes of molecules are known, as well as macroscopic mixtures of different molecules with some intermolecular structure, e.g., certain types of solutions. Additional descriptions of various types of structures composed of various kinds of things (including molecules) can be found in Whyte *et al.* (1969) and in Weiss (1970).

In the Oppenheim-Putnam model the domain of living cells follows the domain of molecules. The preceding discussion indicates at least one respect in which their model is oversimplified. As we have seen, there are a variety of quite different types of things composed of molecules. Moreover, the structures of micelles, for instance, are so different from the structures of, say, macroscopic gas samples, that it is reasonable and convenient to suppose that they are two different genera of structures. These two different genera of structures might then comprise two different domains of two different theories. Thus, at least after the domain of molecules, there will be several different theories whose domains correspond to the different genera of structures composed of molecules. Therefore, the set of domains and their corresponding theories will not be linearly ordered; instead, the array of domains and theories will at least branch at various places. Branching of this kind is no serious obstacle for the unification program. However, it does make the program more complex by requiring successive microreductions down a number of branches in the array of theories.

There is another type of complication resulting from the variety of structures and theories. Sometimes specialized theories are developed in order to account for phenomena involving the interaction of elements from different domains. For instance, if we consider macroscopic samples of gases, liquids, and solids to fall into different domains, we must still account for their interactions and transformations of state. A more serious example is that of solid state physics, which combines aspects of the quantum theory of electrons with certain semi-macroscopic features of crystalline lattices. Another example is provided by the theory of heterogeneous chemical reactions. For instance, some types of gas molecules can be adsorbed onto the surfaces of certain types of macroscopic solids. Chemical reactions may then occur between the adsorbed molecules. There are special laws about the adsorption of layers of gas molecules on solid surfaces. The explanation of such laws may involve relationships between single molecules and crystalline lattices formed of many molecules.

Quite often when it is desired to explain interactions between elements from different domains, it is sufficient to make simultaneous use of the corresponding theories of these domains. In this type of case the unification program would still only require the microreduction of these separate theories. However, more complex situations could arise.

Suppose that we have two theories, T_i, T_j, with disjoint domains Dom_i, Dom_j. Suppose that the basic elements of these two domains can combine together to form a genus of structures identical with the elements in some Dom_k which has a theory T_k. I have described above how the array of domains and theories branches in certain places. I am now describing a possible way in which two domains and theories may converge into a new domain and theory. If such a convergence does occur, it will be necessary to microreduce T_k to the combination of T_i with T_j. This will not be a uniform microreduction exactly like those described in earlier chapters. Nevertheless, this reduction will involve the derivation of the fundamental law-sentences of T_k from the fundamental law-sentences of $T_i \cup T_j$ together with connecting sentences which are identities. Thus the basic kinds of things and the basic attributes of T_k must be identified with things and attributes which are denoted by predicates in, or definable in, the union of the languages of T_i and T_j. Of course, $T_i \cup T_j$ may not be a unified theory. This fact does not interfere with the unification program provided that T_i and T_j can be microreduced to theories with simpler basic elements.

In view of the preceding discussion, it is clear that the unification program must apply to a complex array of many different theories. The simple diagram of Figure 3 indicates the general kind of organization this array of theories might have.

In Figure 3, the dots represent various scientific theories associated with different domains of things. The dots are connected by arrows, which represent relationships between the theories represented by the dots. The basic elements in the domain of a theory at the point of an arrow are structures of a certain genus whose parts are basic elements in the domain of the theory at the end of the arrow. Dot *b* has arrows pointing to several other dots; it is a branching point in the array of theories. Dot *f* illustrates a case of convergence. The basic elements in the domain of the theory represented by *f* are structures composed of basic elements from the

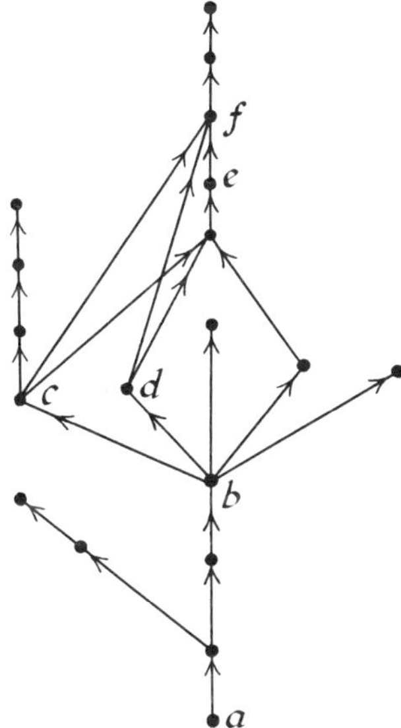

Figure 3. Diagram of a possible array of theories (see text).

domains of the theories represented by *c*, *d*, and *e*. Dot *a* represents whatever theory of fundamental particles is used in the array of theories. This theory of fundamental particles should be a unified theory. Any theory other than theory *a* must be microreduced. A theory at the point of one arrow must be microreduced to the theory at the end of that arrow. A theory at the points of several arrows must be microreduced to the union of the theories at the ends of all of these arrows.

Figure 3 is not intended to be an accurate picture of all present or future theories. This figure merely indicates the general type of organization which may be expected in a systematic array of theories. The program of unification by means of successive microreductions should be understood as applying to an array of theories of the same general form as, but more complex than, that represented by Figure 3.

B. HIERARCHICAL STRUCTURES

The preceding section shows how the unification program is affected by the complexity of the array of scientific theories. This complexity results from the many genera of structures which can comprise the domains of significant scientific theories. In addition, there are certain types of structures which create special difficulties in theory construction and reduction. In particular, in recent years there has been considerable interest in so-called "hierarchical structures"; see, for instance, Pattee (1973), Whyte *et al.* (1969), Weiss (1970), and Weiss (1971). In part this has been interest in problems of the evolution and development of hierarchical structures. These problems will be briefly considered in the next chapter. In addition, there is interest in dynamic theories of hierarchical structures, and in the reduction of such theories.

Unfortunately, there is no generally accepted definition of the notion of hierarchical structure. Often a hierarchical structure is considered to be an entity which is composed of certain kinds of parts, each of which is in turn composed of other kinds of parts, etc. We thus have the whole, and then its parts, and then their sub-parts, and so on, arranged in some kind of order. Organizational structures sometimes have this form. For example, an army might be divided into corps, which are divided into divisions, which are divided into brigades, on down through regiments, battalions, companies, and squads. This organization usually corresponds quite directly with the chain of command. In normal operations, orders from the general thus affect the behavior of the officers and troops on the lower levels of the hierarchy.

Hierarchical structures are also found in organisms. In physiology a multicelled organism is often considered to be composed of systems, which are composed of organs, then tissues, and then cells. Actually, it seems to me that the organization is not quite this simple, for the organism also contains certain substances (e.g., water, hormones) which do not fall into any of these categories and yet are important *parts* of the organism. Thus I will *not* assume that the various kinds of parts of a hierarchical structure must form a linear order of parts, sub-parts, sub-sub-parts, etc. Instead, I will consider a *hierarchical structure* to be one which involves at least two structural descriptions, namely, a description of the whole as composed of certain kinds of parts, plus a description of at least one of these parts as a structure composed of certain sub-parts. As a consequence, the description of a hierarchical structure not only involves at least two distinct structural descriptions, but also at least two distinct genera of parts. Most interesting hierarchical structures are complex and will be associated with complex theories. In the remainder of this section, I will only discuss two aspects of hierarchical structures which might be thought to create serious obstacles for the unification program. Fortunately, these aspects can be illustrated by a simple model of a hypothetical hierarchical structure.

Suppose that there is a certain class of entities known as W's, another class known as P's, and another class known as Q's. Suppose that each kind of W is known to be a certain kind of structure composed of P's, and suppose that each kind of P is known to be a certain kind of structure composed of Q's. If W's, P's, and Q's occur independently of each other in nature and can be studied in their free forms, then it is possible that there would be three separate theories of these three kinds of entities. In this case, the unification program requires that these three theories be successively microreduced. This is the usual situation faced by the unification program.

Now suppose instead that Q's and W's naturally occur in the free form, but that P's do not. Also suppose that P's cannot easily be studied outside of W's. In this case we might construct a theory of the behavior of P's within W's. Such a theory would be somewhat restricted in scope because it would be concerned with the behavior of P's within the restricted boundary conditions imposed by the fact that the P's are parts of W's. If we attempt to microreduce this theory of the P's to the theory of the Q's, we would have to take into account these special boundary conditions. If, in doing this, we make essential reference to W's or special attributes of W's, we would not have a genuine microreduction to the theory of the Q's. This type of consideration has been partly involved in certain arguments for the irreducibility of biology to physics and chemistry; see Polanyi (1968), and, for replies, see Causey (1969b), and Hull (1974, pp. 139–141). However, this type of situation is not a serious obstacle for the unification program. If the theory of the P's utilizes certain special boundary conditions, then the microreducing theory should be the theory of the Q's restricted to these special boundary conditions. In fact, in this type of situation it is probably best not to consider the theory of the P's to be an independent theory in the array of theories to be successively microreduced. Instead, it can be thought of as an intermediate step in the eventual development of an adequate theory of the W's. If a theory of the W's is developed, then the unification program would require reducing it to the theory of the Q's. This reduction would be a two-stage microreduction making use of the structural descriptions of both P's and W's. It would thus be more complex than the single-stage uniform microreductions previously described in detail. However, such a two-stage microreduction is a natural generalization of the single-stage variety.

The complication just discussed arises from the fact that the parts of a structure are under the influence of special boundary conditions created by the structure. This is generally true of all kinds of structures, although it is most apparent in the intermediate levels of a hierarchical structure, especially if we attempt to construct theories of these intermediate levels. This type of theory is also subject to another, related kind of complication.

Again suppose that we have W's, P's, and Q's as above. Suppose that a theory of the P's is developed. It is quite likely that the language of this theory will contain terms which make some reference to W's or attributes of W's. This point can be illustrated by the theory of genes, which was invented in order to explain the hereditary transmission of certain phenotypic characteristics of organisms. Genes were hypothesized as parts of organisms, and these parts were supposed to carry and transmit hereditary characteristics. Eventually a rather complex theory of gene behavior was developed. The domain of this theory is a set of genes of various types. However, this theory also makes reference to phenotypic characteristics of the organisms. For instance, it is supposed that there is a pea plant gene for smooth peas, and a pea plant gene for wrinkled peas. Thus, the classical gene theory makes reference to objects and attributes which are, strictly speaking, not part of the domain of this theory. It therefore does not satisfy the necessary conditions, stated in Chapter 6, for unified theories. Of course, complex organisms involve many structural levels, but the same type of situation could arise in our hypothetical hierarchical structures. A theory of the P's might be developed, and this theory might utilize some predicates which refer to attributes of W's. If this theory of the P's is very important, scientists might try to microreduce it to the theory of the Q's. This could lead to severe complications. Something analogous to this has occurred in modern biology.

It is well known that genes have been identified with small segments of DNA (sometimes RNA) molecules. It is also known that these molecular segments, through complex chemical processes, can help to produce certain protein molecules which in turn help to control complex biochemical processes eventually leading to phenotypic characteristics. Some have been inclined to think that classical genetic theory has, in effect, been microreduced to molecular theory. Others have disagreed and have pointed to the fact that we still know very little about the complex processes between the synthesis of the original proteins and the eventual formation of phenotypic characteristics. The issue has now become quite controversial, for instance, see Schaffner (1974) and Hull (1974, Chapter 1). This controversy largely results from the fact that the theory of genetics refers both to genes and to gross phenotypic characteristics of complex organisms. A similar type of controversy might develop concerning the reducibility of our hypothetical theory of P's which refers to attributes of W's.

It seems to me that there are two reasonable ways the unification program might approach this type of situation. In some cases it might be possible and appropriate to sanitize the theory of the P's by restricting it in such a way that it no longer makes reference to W's or attributes of W's. Thus, the theory of genes could be restricted to the general laws of gene interactions without reference to any particular phenotypic characteristics. If this approach is not acceptable, then we could use the approach suggested in connection with the boundary condition problem. The theory of the P's could be considered an intermediate step in the eventual development of an adequate theory of the W's. In effect, we would admit that the theory of the P's is not *by itself* microreducible to the theory of the Q's. But then we need not, and would not, include the theory of the P's in the array of independent theories to be unified. The theory of the W's would occur in this array, and it would have to be microreduced. If its microreduction could be accomplished, then, *a fortiori*, we would also have microreduced the theory of the P's. This process would be an *indirect reduction* of the P-theory *through the higher-level W-theory*. For further discussion of this procedure see Causey (1969b).

Hierarchical structures obviously create complications in the unification program. However, the dynamic theories of these structures do not appear to be overwhelming obstacles for the program. It is important to bear in mind that the unification program does not require the *direct* microreduction of the theories of intermediate parts of a hierarchical structure if these theories make reference to the whole structure or to higher-level parts of the structure. The following consideration provides a simple way to see this point.

Suppose that we have a linearly ordered hierarchical structure, in which the smallest parts are considered to be the lowest level, and the entire structure is considered to be the highest level. Consider a theory about some intermediate level of parts. Such a theory may use a relational attribute-predicate some of whose relata are higher-level parts of the hierarchical structure. If this is the case, this theory is not a unified theory. Moreover, the level of such a relational attribute-predicate should be considered equal to that of the highest level of parts of the hierarchical structure to which this predicate refers. Thus, the theory in question is really not a well-formulated theory. There is no need to consider it one of the independent theories to be included in the unification program. Of course, we probably would need to microreduce the theory concerned with the entire hierarchical structure. If this is accomplished, then the intermediate theory in question would be *indirectly* reduced through the higher levels, as indicated above.

C. TOKENISM

In several publications, J. A. Fodor and others have presented what are often called "functionalist" arguments against the likelihood of the reducibility of psychology to neurophysiology. These functionalist arguments have already been criticized in the philosophical literature (see Kalke (1969) and Gendron (1971) for criticism and bibliographical information). In Fodor (1974), 'Special Sciences (Or: The Disunity of Science As a Working Hypothesis)', Fodor has extended some of his previous views about psychology into a general discussion of the unity of science. He first criticizes a purely reductionistic program for the unification of science, and then he defends an alternative program. In this section I will first review his criticism of reductionism, then describe his alternative proposal, and finally assess his views. It will be seen that we are not compelled to accept his alternative and, in addition, that this alternative suffers internal problems. Unless otherwise indicated, all page references in this section are to Fodor (1974).

1. Fodor asks us to consider some law-sentence

$$S_1 x \to S_2 x \tag{7.1}$$

of some special science other than physics, where 'S_1' and 'S_2' are not predicates in the language of the basic physical theory, whatever it is. Suppose that all scientific theories are successively microreduced to this basic physical theory. Then, ultimately, 'S_1' and 'S_2' are related by some kind of connecting sentences (which Fodor calls "bridge laws") to predicates defined in the language of the basic physical theory.

Fodor expresses some uncertainty about exactly how law-sentences and natural kinds are to be explicated. However, he writes under the assumption that the natural kind predicates of a branch of science are those predicates which are used in the formulation of the law-sentences of that branch of science. Then, on his view, if all sciences can be successively reduced to physics, then for any natural kind of any science, there is some physical natural kind which is identical with, or co-extensive with, that natural kind. He says that we obtain identity of natural kinds if connecting sentences express property identities, and we obtain co-extensionality of natural kinds if connecting sentences express event identities (p. 102). Of course, in this book I have defended the position that connecting sentences should be thing-identities and attribute-identities. Moreover, I do not see why Fodor should consider the co-extensionalities to be event identities, and I disagree with him that the co-extensionalities should be nomological (p. 104). I have argued at length in earlier chapters that identities are not nomological. Nevertheless, these issues can be ignored at this point, for they are somewhat independent of Fodor's main reasons for doubting the wisdom of the reductionistic program for the unification of science. Fodor simply doubts that, in general, for any natural kind predicate in any scientific theory there will be a physical natural kind predicate co-extensional with that predicate.

Fodor admits that the success or failure of the reductionistic program is ultimately an empirical question, and therefore he realizes that his reasons for doubting this program are largely empirical speculations. Nevertheless, they seem to have some force, especially in view of the two examples which I will now describe.

The first example is concerned with economics. Fodor supposes that economic theory will have a predicate, "monetary exchange", and he is willing to believe that any event which is a monetary exchange has a true description in physical language such that the laws of physics apply to that description. However, it appears that, in various contexts, many different physical descriptions would describe events which are monetary exchanges. Therefore, he doubts that there could be a physical natural kind predicate related by a co-extensionality connecting sentence to "monetary exchange" (p. 103). Furthermore, even if some physical predicate were co-extensional with "monetary exchange", he apparently believes that this co-extensionality would most likely be an accidental co-extensionality (p. 104).

His second example is similar, but is concerned with psychology. He considers psychological theory to be a very general theory which applies to various kinds of organisms, and possibly also to certain types of automata. In such a theory a psychological predicate denotes a certain type of psychological state. Under various conditions, the same kind of psychological state may occur in various different organisms and automata. Yet, these organisms and automata may have very different internal structures and states. Therefore, Fodor believes that it is unlikely that there could be a physical natural kind predicate related by a co-extensionality connecting sentence to a psychological natural kind predicate. However, even if there are no such connecting sentences, he believes that it is still possible that each particular psychological event could be identical with some particular physical event (pp. 105–107).

Fodor's views about the reducibility of psychology have been criticized at length by Kalke (1969) and Gendron (1971). Their critiques cast serious doubt on the ultimate significance of his views. At this point I will not directly reply to the two examples just described. Instead, I will now discuss Fodor's alternative program for the unification of science. Examination of this program will facilitate further consideration of his two examples.

2. Consider (7.1), and suppose, with Fodor, that 'S_1' and 'S_2' are natural kind predicates of some special, nonphysical science (e.g., psychology). Thus, according to Fodor, these predicates refer in some way to natural kinds of this special science. Also, as we saw above, he believes that it is in general unlikely that these natural kinds are identical with, or nomologically co-extensional with, physical natural kinds. Thus, he believes that the special science is not reducible to physics by the use of connecting sentences which express identities or nomological co-extensionalities. He therefore considers a much weaker type of connecting sentence of the form

$$S_1 x \rightleftarrows P_1 x \vee P_2 x \vee \cdots \vee P_n x, \tag{7.2}$$

in which 'S_1' is a natural kind predicate of a nonphysical science and the 'P_i' are physical predicates. He expects that, in general, the entire disjunction on the right side of (7.2) will not be a physical natural kind. The exact interpretation of the '$\rightleftarrows$' is not entirely clear. However, (7.2) is supposed to state a true generalization. Also (7.2) is supposed to be interpreted as indicating some kind of "event identities". In particular, any "event" which is an instantiation of S_1 is supposed to be identical, according to (7.2), to some event which is an instantiation of one of the P_i. Finally, (7.2) will not, in general, be a nomological co-extensionality because in general the disjunction on the right side of (7.2) is not a natural kind predicate (p. 108).

In Fodor's view (7.1) is not likely to be reducible (in any normal sense of "reduction") to a physical law-sentence. Fodor believes this because he believes that it is unlikely that S_1 and S_2 are identical with, or nomologically co-extensional with, physical natural kinds. Suppose that this is the case, and suppose also that we have (7.2) and

$$S_2 x \rightleftarrows Q_1 x \vee Q_2 x \vee \cdots \vee Q_m x, \tag{7.3}$$

where the 'Q_j' are physical predicates and (7.3) expresses the same kind of relationship as that expressed by (7.2). Then (7.1) corresponds in physics to

$$P_1 x \vee \cdots \vee P_n x \xrightarrow{w.e.} Q_1 x \vee \cdots \vee Q_m x, \tag{7.4}$$

where '$\xrightarrow{w.e.}$' means 'if, then, with exceptions'. This peculiar '$\xrightarrow{w.e.}$' notation is mine, not Fodor's, but it facilitates the exposition of his view.

Recall that (7.2) and (7.3) are supposed to be true generalizations expressing certain types of event identities. Thus, an event which is an instantiation of S_1 is supposed to be identical with an instantiation of one of the P_i, and similarly for S_2 and the Q_j. Yet Fodor believes that the laws of nonphysical sciences usually have exceptions. Suppose that this is the case with (7.1). Fodor interprets this in the following way. There may be some S_1 event, which by (7.2) is identical with some P_i event, such that this S_1 event is not associated with any S_2 event. According to Fodor, this implies that the P_i event is not nomologically connected with any Q_j event. Thus we have an exception to (7.1). Of course, this implies that (7.4) has exceptions, whence the notation '$\xrightarrow{w.e.}$'. Needless to say, (7.4) is not a law-sentence when it has such exceptions. Fodor believes that we could *impose* the natural kinds of physics upon the special, nonphysical sciences. But he thinks that we should not, and will not, do this. Instead the special sciences will be allowed to use their own special natural kind predicates, and will continue to have laws with exceptions. These exceptions are to be "explained away" within the physical sciences (p. 112). Fodor believes that his discussion shows why there are special sciences which refer to natural kinds which are not identical with, nor nomologically co-extensional with, the natural kinds of physics.

I believe that I can now give a fair summary of Fodor's alternative program for the unification of science. In his view there are many different branches of science with which are associated different theories. Physics (in some form) is probably the basic science. Some other theories may be reducible to physics by using connecting sentences which state natural kind identities, or nomological co-extensionalities of natural kinds. However, there are also special branches of science which are not reducible in this manner to physics. These special branches of science use natural kind predicates which are not co-extensional with physical natural kind predicates. These special branches of science also state laws which have exceptions. Science cannot be unified by the reduction of such branches of science to physics, but reduction is not necessary for the unification of science. All that is necessary for unification is that all *things* and all *events* be things and events of the basic science. In particular, if physics is the basic science, then any event within the domain of any branch of science must be shown to be a physical event. The reduction of this branch of science will show this, but reduction is not the only way to show it. We can also have token event identities expressed by sentences such as (7.2) and (7.3). Fodor gives the name "token physicalism" to the claim that any event in the domain of any branch of science is a physical event (p. 100). Fodor's program for the unification of science amounts to showing that physics is the general, basic science, and that token physicalism is true. In part this may be accomplished by reductions; but its accomplishment will probably also require some connections such as (7.2) and (7.3). Thus, there can be unity of science by means of token physicalism, together with the existence of special sciences which are not reducible to physics. In order to have a short name for Fodor's program for the unification of science, I will call it 'tokenism'.

3. There are a great number of comments one could make about Fodor's doubts concerning reductions, and about his tokenism. However, it is clear that his attitudes towards both reductionism and tokenism depend heavily on his views concerning laws and naturals kinds. I will

therefore limit the present discussion to three comments about his views on laws and natural kinds.

First of all, recall Fodor's example from psychology. He believes that a particular natural kind in psychology may correspond in different circumstances to many different kinds of physical states. However, he does not present any definite criteria which would determine what counts as a natural kind predicate in psychology. For illustrative purposes, he suggests that we might identify psychological event types by their behavioral consequences (p. 106). However, on p. 115, fn. 5, he rejects this suggestion and refers instead to Block and Fodor (1972) for what he considers to be a more reasonable criterion. Now in this latter article Block and Fodor do not totally commit themselves to a criterion of identity, but they are at least sympathetic towards the "... doctrine which holds that the type-identity conditions for psychological states refer only to their relations to inputs, outputs, and one another" (Block and Fodor (1972, p. 173)). This is still an extremely liberal criterion for the identification of types of psychological states, and Block and Fodor state some particular reasons for doubting its adequacy. Moreover, it is not even consistent with another criterion they suggest elsewhere in their article. On p. 180 Block and Fodor suggest that the identity criteria for psychological states will depend on the psychological, and perhaps neurological, laws involving these states. Thus they suggest two different criteria; the first is based on a kind of purely psychological theory, while the second also takes into account neurological laws involving psychological states. If one uses the first criterion to individuate types of psychological states and events, then one will naturally be doubtful of the reducibility of psychology to neurophysiology, for one might find "the same psychological state" correlated with different physiological states. Thus one would also naturally be sympathetic towards tokenism. On the other hand, if one uses the second criterion, then it is not at all obvious that one should favor tokenism over reductionism. Consistent use of the second criterion might lead us to split "the same state", according to the first criterion, into many different types of states. I have, in effect, made this observation previously in Causey (1976b).

In view of what was said in Chapter 2 about laws and identities, it should be clear that it is risky to argue for or against certain attribute-identities without clear and stringent criteria for such identities. Moreover, as was made clear in that chapter, the individuation of kinds and attributes is interdependent with the individuation of laws and explanations. Fodor seems convinced that we will always individuate psychological states in a manner which does not correspond well with our individuation of physical states. However, he has not stated and defended definite identity criteria according to which this must be the case. The force of his example from psychology is therefore considerably weakened. Likewise, the impulse towards tokenism is greatly retarded.

My second comment is also concerned with Fodor's views about natural kinds and events. In (7.2) the natural kind predicate 'S_1' is related to a disjunction of physical predicates. Any event which is an instantiation of S_1 is supposed to be token-identical with some physical event, yet S_1 is supposed not to be co-extensional with any physical natural kind. Fodor argues (pp. 106–107) that we could have evidence for the truth of the token event identities without there being co-extensionalities of natural kinds. This argument is formulated within the context of his example from psychology. But the kind of argument that he presents assumes, in effect, that we will maintain the belief that 'S_1' denotes a single natural kind, while we discover that instantiations of S_1 are sometimes instantiations of P_1, sometimes instantiations of P_2, etc. How do we justify these discoveries? Fodor believes that this can be done if the physical theory can explain why in some situations an instantiation of P_1 behaves as an instantiation of S_1, why in some situations an instantiation of P_2 behaves as an instantiation of S_1, etc.

Now clearly, if we are going to justify (7.2) in this manner, we cannot hope to provide such explanations for each *particular* event which is an instantiation of P_1, P_2, etc. Most scientific explanations, especially those in a general theoretical context, apply to *kinds* of events, not particular, individual events. Therefore, in order to justify (7.2) in the manner proposed by Fodor, we must be able to justify why, in circumstances C_1, say, S_1 is co-extensional with P_1, and why, in circumstances C_2, S_1 is co-extensional with P_2, etc. But if all of this can be done, it is certainly not obvious that we would continue to consider S_1 as a single natural kind. Several alternative possibilities might appear more reasonable depending upon other aspects of the theoretical situation.

In some cases we might decide to split S_1 into n different natural kinds which we believe we can justify as being identical with $P_1, P_2, \ldots, P_n$, respectively. In other cases we may have causal explanations of the co-extensionalities. Then the various co-extensionalities would be nomological. If the C_i and the P_i are sufficiently different in kind (e.g., some may be electrical, some mechanical, some chemical), we may still decide to split S_1 into different natural kinds corresponding to the various theoretical domains of the C_i and P_i. Finally, if the C_i and P_i are all included within one theoretical domain, they may have some more general, common aspect which can be identified or correlated with S_1. In this case S_1 would still be considered to be one natural kind, but also the disjunction in (7.2) would probably be considered to refer to one natural kind.

The point of all of this discussion is very simple. At first glance tokenism appears to be a relatively modest goal which should be considerably easier to accomplish than unification by successive micro-reductions. Yet, when we consider in detail what is involved in justifying sentences such as (7.2), it becomes clear that accomplishing tokenism really involves a great deal. Moreover, establishing sentences such as (7.2) actually involves doing many of the same things we must do in the reductionistic program. Finally, when these things are done, they may provide good reasons for concluding that S_1 is not a single natural kind, or that the disjunctions in sentences like (7.2) are natural kinds. Indeed, attempts to carry out tokenism may lead us to modify the languages of the special sciences. I have previously discussed such modifications in Chapter 5, Section E. In view of all of this, it would be gratuitous to maintain that we will always use the natural kinds of the special sciences, and that therefore tokenism is a more appropriate program for unification than reductionism.

My final comment is concerned with Fodor's interpretation of the laws of the nonphysical sciences. Fodor gives no general characterization of what counts as a law or as a natural kind. He merely assumes that the natural kind predicates of a science are those predicates which are used in the formulation of the law-sentences of that science. Yet, as we have seen, Fodor makes many strong claims about laws and natural kinds. Among other things, he seems confident that the laws of the basic science (probably physics) should not have exceptions, and he is also convinced that the laws of the special, nonphysical sciences do have exceptions. For instance, (7.1) is supposed to have exceptions. These exceptions are reflected in (7.4), which I symbolized with '$\xrightarrow[w.e.]{}$'. According to Fodor, there may be an S_1 event which is identical with some P_i event which is not nomologically connected with any Q_j event. Fodor admits that we could reformulate the nonphysical law-sentences so that they would then be exceptionless by *imposing* physical natural kind predicates upon the nonphysical sciences. But he thinks that this will not, and should not, be done. Instead we will keep nonphysical law-sentences such as (7.1), keep the nonphysical natural kinds, keep the exceptions, and then "explain away" the exceptions in terms of the basic physics. In view of the fact that he presents no general characterization of laws and natural kinds, it is hard to see why he is so confident about what we should, and will, do.

In Chapters 2 and 3 I presented extensive discussions of law-sentences, thing-predicates, attribute-predicates, and classification systems. At the very least these discussions should have made it clear that the construction of a scientific language and its classification scheme goes hand-in-hand with the development of a theory of the domain with which this language is concerned. The language of a theory is intimately connected with the law-sentences of the theory. The discovery of a new kind of thing or a new attribute may lead us to modify previously accepted laws. The discovery of new laws, or of exceptions to an old law, may lead us to modify our theoretical language. Chapter 5, Section E, presents examples of such modifications. In addition, there is the frequently mentioned example of whales. It is likely that once upon a time some people included *whales* in the natural kind, *fishes*. After gaining more knowledge about fishes and other animals, *whales* came to be classified as *mammals*. These general observations should at least make one skeptical of Fodor's belief that we will retain nonphysical laws with exceptions. A closer consideration of his discussion should increase this skepticism.

Suppose that we have (7.1) through (7.4), and suppose that we discover that instantiations of P_1 are not nomologically connected with any Q_j. If no instantiation of P_1 is ever associated with an instantiation of any Q_j, then we would have good reasons for modifying S_1. We would probably redefine 'S_1' in such a way that circumstances satisfying 'P_1' are no longer considered to be circumstances satisfying 'S_1'.

However, it might happen that some instantiations of P_1 are associated with instantiations of some of the Q_j, and others are not. Then we would wish to discover and explain the physical laws governing the various kinds of relationships between P_1, boundary conditions, and the Q_j. If this is done, it might be said that we have "explained away" some exceptions to (7.1). But in fact more has been accomplished. We are now in a position to modify both S_1 and (7.2) in such a way that these "exceptions" are eliminated. Of course, neither Fodor nor I can prove what scientists will do in such hypothetical situations. However, I believe that it is more likely than not that they would remove the exceptions by performing the appropriate modifications. In general, when we discover exceptions to old law-sentences, we try to reformulate them in such a way that the reformulated law-sentences do not have exceptions.

The first of my three comments considerably weakens Fodor's reasons for doubting the prospects of the reductionistic program for the unification of science. My second and third comments raise serious questions about Fodor's conception of natural kinds, event identities, and laws. My second and third comments therefore further undermine Fodor's reasons for preferring tokenism, and they raise doubts about the methodological wisdom of tokenism. Incidentally, the second comment also shows that accomplishing tokenism requires much of the same kind of scientific work that is involved in the reductionistic program. I conclude that the program for the unification of science by successive microreductions is a more viable and methodologically sound program than tokenism.

Needless to say, I have still not directly discussed the difficulty caused by Fodor's example of 'monetary exchange'. Since this example raises special problems about the reduction of the social sciences, I will discuss it in the next section.

D. SOCIAL THEORIES AND SOCIAL STRUCTURES

The social sciences are primarily concerned with the collective behavior of various kinds of groups, and secondarily with the behavior of individuals within groups. Traditionally, the individuals are human beings, although we could consider the social sciences to include the study of the behavior of groups of animals. The main fields of human social science are sociology, political science, economics, anthropology, history, and social psychology. Since the unification program applies directly to dynamic theories, historical studies will not be of immediate concern in this section. Chapter 8 contains some discussion of historical knowledge and explanation.

In comparison with the physical sciences, the social sciences are at present relatively undeveloped. This is not surprising since the latter are largely concerned with extremely complex systems which can exhibit an enormous variety of types of behavior. It is clearly risky to speculate about the development of dynamic social theories, and, in particular, about their reducibility to theories about individual behavior. But social structures and social theories occupy important positions in the total array of structures and theories. It is necessary to provide at least a sketchy indication of how they can be fitted into the unification program.

We usually think of a human social group as a collection of individual people interacting in various ways. Some groups, for example an army, are large and involve complex social relationships. Others, for example, a few customers in a café, are small and may be relatively unstructured. In general, I consider a human social group to consist of a set of two or more people satisfying some social structural description. Since there are many different general categories of social structures, it is natural that there are different fields of social science concerned with these various categories. I will therefore let T_2 be the dynamic theory of some field of social science concerned with some general category of structures. This section is concerned with some general features of such theories, and with some problems involved in the reduction of such theories to some theory, T_1, of individual psychology. The topic of this section is therefore closely related to the thesis of methodological individualism, see, for instance, Brodbeck (1958).

Now it is a widely accepted fact that all people, with the exception of hermits and island castaways, live under the influence of some social structures. Consequently explanations of most behavior of individuals will make or presuppose some reference to social structures or structural relationships. An individual's behavior in a society is under the continual influence of peer group relationships, prescriptive laws, social conventions, economic constraints, etc. Therefore the behavior of an individual in a society will in general *not* be *directly* describable in terms of $\mathscr{L}_1$, the language of T_1. Similarly, explanations of his behavior will in general have to take account of the relevant social setting. Of course, if T_2 is microreduced to T_1, then ultimately his behavior would be describable in $\mathscr{L}_1$, and explainable in terms of the laws of T_1 plus the relevant identities and boundary conditions. Needless to say, this reduction requires that the relevant social structures and social attributes be describable in $\mathscr{L}_1$. These simple observations seem often to be forgotten in discussions about the explanation of behavior. Yet, they are elementary observations which have direct analogs in other contexts. For example, if we wish to explain the behavior of a particular kind of atom within a particular kind of molecular structure, the boundary conditions of this explanation must take into account the relevant structural relationships between this kind of atom and the others in the molecular structure.

In practice, it is much easier to study the behavior of uncombined atoms than it is to study the behavior of people freed from social relationships. Yet, in principle, I see no reason why a significant theory of individual psychology could not be constructed and confirmed. This theory would contain laws about very basic types of behavior and psychological states. Among other things, it would have to include laws about internal motivational states, perceptual processes, memory, belief states, emotions, and voluntary movements. At the present time we are only beginning to learn some of the laws governing these kinds of states and processes. But let us suppose that such a theory T_1 can be developed. Then $\mathscr{L}_1$ should enable us to describe attitudes and beliefs of one individual about other individuals and groups of individuals.

Let us now consider T_2 and $\mathscr{L}_2$. T_2 is a theory about a certain category of social groups. Within this category there may be many different possible types of groups with the same *very general* structural features. T_2 might be a theory about friendship cliques or a theory about industrial firms, or some more general theory that contains both of these categories as subcategories of some much more general category. If T_2 is a very general theory of sociology, it may deal with complex hierarchical structures composed of groups, subgroups, etc. Since the special complications of hierarchical structures are discussed in an earlier section, I will simplify the present discussion by supposing that Dom_2 consists of groups whose parts are individual persons. Then $\mathscr{L}_2$ has thing-predicates for the various kinds of groups in Dom_2, and attribute-predicates for the attributes of these groups. T_2 contains law-sentences about the attributes and behavior of these groups. For example, one law-sentence of T_2 may relate the *cohesiveness* of certain types of groups to some other attribute of these groups. Another law-sentence may be concerned with the relative efficiency of certain types of groups in the performance of certain types of tasks. Whatever may be the specific subject matter of T_2, it must be remembered that T_2 is a theory about groups and the attributes and behavior of groups.

In order to reduce T_2 to T_1 it is necessary to identify the thing- and attribute-predicates of $\mathscr{L}_2$ with thing- and attribute-predicates in, or defined in, $\mathscr{L}_1$, With the help of these identities, the fundamental law-sentences of T_2 must be explained in terms of law-sentences of T_1. Therefore, as is the case in all microreductions, T_1 must be a theory with structured wholes. In particular, the various kinds of groups in Dom_2 must be denoted by structural descriptions formulated in $\mathscr{L}_1$.

Much of current social science is concerned with the formulation of adequate descriptions of various types of social structures. This work can be considered a first, partial step in the direction of an eventual reduction of sociology to individual psychology. Of course, current social science still uses many terms not found in the still developing theory of individual psychology. Many organized social structures are presently described in terms of statuses, relationships between roles, associational norms, etc. On the other hand, some relatively simple structures, such as friendship cliques, can be described using sociograms with simple binary relationships between individuals. These simple binary relationships, e.g., person a likes person b and person b likes person a, would be relatively easy to define within a significant $\mathscr{L}_1$.

The overall strategy of the reduction should now be clear. $\mathscr{L}_1$ will probably contain some relatively simple structural relation-predicates (see Chapter 3, Section D). These will make possible the descriptions of certain simple types of structures. In addition, more complex structural descriptions will be constructed using these, and other predicates of $\mathscr{L}_1$. A social structure which involves statuses and associational norms will probably be very difficult to describe in $\mathscr{L}_1$. Consider an American football team. There are many different playing positions on the team. Each position has certain roles to play and, in addition, there are many complex rules governing scoring, penalties, and timing of the game. Could we construct, in $\mathscr{L}_1$, a description of the structure of a football team? Remember that this description must be identified with the usual description given in terms of a language which refers to statuses, norms, and rules of play.

First of all, in order to have a team, it is necessary to have persons who know the various team positions and the roles of these positions. These persons must know their own positions and especially their own roles. It is conceivable that these conditions could be describable in terms of certain cognitive and intentional states and certain behavioral dispositions. In addition, the players must know certain rules of play, and must know how to react to certain behavior of others. These conditions amount to further cognitive states and behavioral dispositions. No doubt there are additional factors to be taken into consideration. However, it appears at least possible that a significant language, $\mathscr{L}_1$, of individual psychology, should be able to provide the appropriate descriptions of the team structure. Unfortunately, at the present time we are in no position to attempt the task because we do not have an adequate $\mathscr{L}_1$.

If someday an adequate $\mathscr{L}_1$ is developed, there are still further conditions to be satisfied by T_1. First, T_1 must have law-sentences which enable us to explain why certain types of social structures can or cannot exist under specified conditions. Moreover, $\mathscr{L}_1$ must provide descriptions of attributes of social structures, and these descriptions must be identified with attribute-predicates of $\mathscr{L}_2$. Finally, the law-sentences of T_2 must be microreductively explained in terms of the law-sentences of T_1.

One possible objection to this program for reduction can be stated as follows: The rules for football are not natural laws, they are prescriptive conventions. Therefore, these rules cannot be derived from the natural laws of T_1. Yet these rules are involved in a description of the structure of the team. Therefore, it would appear that T_1 could not accomplish one of its necessary tasks, namely, to explain why football teams can or cannot exist under certain conditions.

This objection is based on a confusion. It is true that the rules are conventional, and they undergo changes with the passage of time. In order to account for the origin of particular rules and particular structures, it is necessary to provide historical explanations. T_1 alone is not required to do this. T_1 is required to explain the attributes and behavior of structures, *given* the existence of these structures. T_1 is also required to explain the conditions for the existence of structures, and the objection gives no reason why T_1 cannot do this. For example, some conceivable rules might make effective play psychologically impossible. T_1 should be able to explain such a fact, just as physics could explain why some conceivable rules might make play physically impossible. In short, the conventional aspect of whatever rules might be used is taken account of in $\mathscr{L}_1$ and T_1 by means of cognitive states of the players. Suppose that we have specified a set of rules according to which play is physically possible. Then $\mathscr{L}_1$ should be able to describe the psychological states of a set of people such that these states are necessary and sufficient conditions for the existence of a team which plays according to these rules. T_1 should be able to explain why these conditions are necessary and sufficient for the existence of such a team. On the other hand, the fact that particular rules are used by particular people at a particular time is an historical fact and cannot be explained solely by a dynamic theory alone.

It should now be more clear why, in most practical contexts, the explanation of a particular item of behavior of a particular individual cannot ignore the social context of this individual. The person in question will usually be a member of several different social groups. His attitudes, beliefs, and dispositions will in part depend upon his social context. In order to explain a particular behavior which he produces at a particular time, we will usually need to use several kinds of information. This information may include: laws of T_1; his individual personality traits; his social attitudes, beliefs, and dispositions; further details about the relevant states of his social context, and perhaps certain physical and biological constraints. It will be noted that structural features and other attributes of his social context function as some of the boundary conditions in the explanation of his behavior in terms of the laws of the psychological theory T_1. In most practical contexts explanations of behavior will use a language which is a combination of psychological and social terminology. Social terminology is most likely to be used in making references to statuses, norms, and organizational structures. It may *appear* impossible to explain such behavior without explicit use of social language and social theories. However, if the relevant T_2 and $\mathscr{L}_2$ are reducible to T_1 and $\mathscr{L}_1$, then in principle the behavior is explainable within $\mathscr{L}_1$ using the psychological laws of T_1. This does *not* mean that the social context is eliminated. The social context is still present in the explanation, but it will have been *reduced* to, and hence constructed from, the language and theory of individual psychology.

There is still a final, apparent obstacle which I wish to consider in connection with the social sciences. It will be recalled from the previous section that Fodor based some criticisms of the unification program on some considerations about natural kinds. I have already replied to his general arguments in that previous section. I must now discuss the problem he raises about the predicate, 'x is a monetary exchange'. His reasoning seems to be as follows in Fodor (1974, pp. 103–104): Suppose that this predicate is a natural kind predicate of economic theory, and suppose that it occurs in some law-sentence of this theory. Suppose that economics is reducible to physical theory. Then this predicate must be identical with, or nomologically co-extensional with, some natural kind predicate of physical theory. But it is extremely unlikely that this is the case, because an enormous, heterogeneous variety of physical events would correspond to events which are monetary exchanges. Even if the economic predicate is co-extensional with some physical predicate, Fodor believes that this would be an accidental co-extensionality and that the physical predicate would probably not be a natural kind predicate. Needless to say, Fodor draws the conclusion that economics is not at all likely to be microreducible to physical theory.

It will be noticed that 'monetary exchange' is definable in terms of 'transaction' and 'money'. Moreover, the general type of argument that Fodor presents could be formulated using the term 'money' instead of 'monetary exchange'. Therefore, I will discuss the shorter term, 'money'.

Now it is well-known that money can be almost anything which can be used conveniently as a standard of value and as a medium of exchange. Therefore, in different contexts, money can correspond to a variety of different physical things. But one cannot simply identify a shell, or a piece of paper, or any other physical thing with money. In order to be money, a kind of thing must play a certain role in a certain social context. This in turn requires that the people in this context have certain dispositions and beliefs. Therefore, an economic law-sentence which uses the term 'money' must be considered to be an abbreviation for a much more complex law-sentence which is, at least in part, about the behavior of people in certain social contexts. The reduction of such a law cannot be accomplished without the reduction of a large part of this entire social context. It is obvious that, in the reduction of this law, the term 'money' cannot simply be replaced by a disjunction of predicates denoting physical objects which are actually used as money. A similar remark applies to the term 'monetary exchange' and the simple physical events which are associated with (but not identical with) monetary exchanges. If Fodor believes that microreductions require such simple replacements, then his argument is based on an incorrect conception of the requirements for microreductions and the argument is invalid. In fact, it is highly unlikely that we will ever directly identify a physical predicate with the term 'money', and the unification program does not require this. This program requires *successive* microreductions of various theories. Thus, we should be concerned with the microreduction of the social sciences to individual psychology, and the latter to neurophysiology, etc. Fodor's discussion of monetary exchanges gives no reason why, in principle, this should not be possible.

Yet, Fodor's discussion *appears* to present a serious obstacle to the unification program. It is clear that the unification program implies that 'monetary exchange' and 'money' should, in principle, be identified with physical predicates. His discussion makes it appear that this requires the simple types of replacements mentioned above. Since these replacements are obviously inadequate and unfeasible, the reduction program *appears* unfeasible. But this appearance is false because the above-mentioned replacements are not even correct identifications and could not be used in adequate microreductions. Instead, the reduction will have to proceed along the following lines.

Economic theory is a kind of social theory. The term 'money' is a social attribute-predicate which has uses within certain types of social systems. The general structure of such systems must be characterized. Also the general norms governing the use of the term 'money' in such systems must be characterized. Economic laws about money can then be stated. These laws are special laws applying to certain types of social structures. They must then be reduced to the psychological theory T_1. All of this is difficult, but it nowhere requires identifying 'money' with a huge disjunction of predicates denoting the physical objects which might be used as money in various systems.

Within a particular social system the term 'money' may apply to certain types of objects according to particular norms of this system. This in turn means that the people in this system have certain specific types of beliefs and dispositions. If a particular society uses a particular kind of object, e.g., shells, for money, this fact will be included in their special norms, and will be reduced by the reduction of these special norms. However, one must give a separate, historical explanation of why they came to have these special norms and use shells for money.

By now it should be quite obvious that microreductions of social theories will involve enormous complexities. It is not surprising that there are many who have believed that such microreductions are impossible. Yet, the reasons that have been given for the alleged impossibility are sometimes based on a failure to recognize the enormous complexity of social phenomena. This *is* perhaps somewhat surprising. It is very easy to describe certain social behavior as "rule governed behavior". Then it is noted that the rules are conventional and that particular rules cannot be derived from general, dynamic laws of psychology. This much is correct. The danger lies in stopping the analysis at this point, and concluding that some social phenomena are therefore irreducible to individual psychology. As we have seen above, this conclusion is unwarranted.

Adequate microreductions of social theories cannot be accomplished in the foreseeable future. Yet, partial results in this direction can be hoped for much sooner, and the unification program has a valuable methodological aspect. This program will require us to describe social structures in terms of the individuals, and the psychological states of the individuals, which compose these structures. This is a first necessary step towards the understanding of why social structures have the properties they have, and also why individuals behave as they do within social structures.

CHAPTER 8

SCIENTIFIC PROGRESS AND UNITY OF SCIENCE

In Section A I describe some general aspects of scientific progress and discuss the role of the unification program in the progress of science. This discussion is concerned with dynamic theories, but scientific progress also involves our knowledge and understanding of developmental and evolutionary processes. Section B describes some important relationships between the unification program and our explanations of these processes. These first two sections outline some general motivations for the unification program. Section C concludes the book with a few remarks on some further problems and prospects.

A. General aspects of scientific progress

In discussions of science and technology, it is customary to distinguish basic research, applied research, and development research. All three categories are often collectively called "R & D". I am inclined to distinguish the three categories functionally in terms of their aims. *Basic research* is an inquiry to acquire knowledge and understanding for no specified purpose other than to increase our knowledge and understanding. *Applied research* is an inquiry to acquire knowledge or understanding for some specified purpose(s) other than simply the acquisition of knowledge or understanding. *Developmental research* may be defined very generally as any rational effort devoted to the creation, fabrication, or design of a specific process, instrument, tool, machine, procedure, technique, etc., for some specified purpose(s) other than simply the acquisition of knowledge or understanding. It is possible to quibble over these definitions. However, anyone seriously interested in scientific and technological progress must take into account the kinds of things and activities to which they refer.

It will be noticed that each of the three definitions involves some kind of goal. Thus, any specific research project will have a target with respect to which the success or failure of that project can be evaluated. In the cases of applied research and developmental research a particular project will have a more or less specific practical aim. In the case of a particular basic research project the aim will be to acquire knowledge or understanding in some specified area of science. It is important to note that a particular research or development project may fail to achieve its specified aim and yet accomplish something else of importance. General scientific and technological progress cannot be measured simply on the basis of whether or not particular projects accomplish their specified aims.

It should be clear from the above discussion that there are many aspects of general scientific and technological progress. However, in this book I have been concerned with knowledge and understanding. Therefore, unless otherwise specified, in the following *scientific progress* will refer to increases in knowledge or understanding. It will not matter what type of research leads to these increases, although many of the most dramatic increases will result from basic research.

As is well known, philosophers have been concerned with the nature of knowledge for over 2500 years. In more recent times there has also been much interest in the nature of scientific explanations and theories. Many kinds of skeptical arguments have been presented which cast doubt on the certainty of our knowledge and the adequacy of our scientific explanations. I will ignore these skeptical arguments here. Needless to say, any scientific beliefs which are to be considered scientific knowledge must be well-confirmed. An important branch of scientific methodology is concerned with the procedures and standards for adequate confirmation. There are many theories of confirmation available in the literature, and there is no need to discuss them here. On the other hand, this book does contain a great deal of discussion of the structure of certain types of theories, and of the adequacy of certain types of explanations. Any *particular* scientific theory or explanation might be thought to contain empirical falsehoods. Thus, particular theories and explanans must pass certain observational tests. Again, I am not here concerned with the reasons for doubting the adequacy of standard scientific testing procedures.

At any given time in a particular branch of science, there is a stock of well-confirmed and generally accepted beliefs which I will call the 'knowledge' of the field at that time. Some of this knowledge may be fairly specific, such as the fact that the gas constant has the value $8.31696 \pm 0.00034 \times 10^7$ erg mole^{-1} deg^{-1}. Some of this knowledge may consist of well-confirmed, very general laws, such as Newton's law of gravity. A branch of science may even contain a unified theory based on several well-confirmed, fundamental laws. In part, scientific progress involves increases in our stock of scientific knowledge. However, such progress cannot be measured simply in terms of the *quantity* of accumulated facts. The *quality* and *informational content* of knowledge is very important.

A great deal of research consists of rather routine fact gathering; most of this work is interesting and valuable. I recall a magazine advertisement by an aluminum company. This advertisement referred to 2,289 man-years of experience with aluminum, and it contained a picture of a man sitting on what was presumably a library of aluminum lore. There is no doubt that much of this aluminum information is interesting and useful. Nevertheless, sometimes a single important discovery can represent far more scientific progress than the accumulation of an entire library of specialized factual information.

As a general rule, in basic research we are most interested in *surprising* new discoveries. For instance, it might be much more important to discover a new virus than to measure the size of a known virus with a little more accuracy than is presently available. Of course, surprising discoveries occasionally result from increasing the accuracy of measurements. Such measurements contributed to Rayleigh's and Ramsay's discovery of inert gases in the atmosphere. However, Rayleigh's measurements were not undertaken simply for the sake of greater accuracy. He originally set out to test Prout's hypothesis that the atomic weights are integral multiples of the atomic weight of hydrogen. This and other information about Rayleigh can be found in Howard (1964).

Very generally, surprising discoveries are those which most scientists do not expect on the basis of current knowledge and understanding. These surprises have a double value: they add to our knowledge and they also exhibit gaps in our understanding. The gaps may be more or less important depending on the circumstances. In some cases, a surprising discovery may be explainable in terms of current scientific knowledge. The development of such explanations can be important. In some cases, a surprising discovery may provide unexpected confirmation or disconfirmation of some current, but still questionable, hypotheses. In some cases, a surprising discovery may turn out to be unexplainable in terms of current knowledge or hypotheses. It may then provide an exceptional challenge to the theoreticians, and its explanation will require the invention of new hypotheses. In extreme cases, a surprising discovery may even conflict with long accepted "knowledge" and thus show that scientists have been in error. Naturally such errors must be corrected by suitable revisions in our stock of beliefs. However, I do not believe that they are justification for extreme skepticism about scientific knowledge.

It would be an interesting problem to undertake a systematic classification and analysis of various kinds of discoveries and their informational content with respect to current knowledge. The above remarks about surprising discoveries are only an indication of some of their characteristics. For a little more discussion of their importance, one may consult Causey (1968b), on which some of the present section is based. My aim here is simply to indicate why some discoveries of new knowledge have greater significance than others. One way science progresses is through the discovery of new knowledge, but the rate of this kind of progress depends on the significance, as well as on the quantity, of this knowledge. At present there is no accepted way to measure the degrees of this significance, but the above remarks show that the significance of a discovery depends on the relationship between this discovery and our current scientific understanding.

Unfortunately, the notion of *understanding* is more difficult to explicate than that of knowledge. It is clear that it is possible to have considerable knowledge and still have very little understanding. It was known for years that bats can fly in total darkness without constantly bumping into objects, yet this fact was very mysterious and not understood. In order to understand how bats can fly in darkness successfully it is necessary to have an *explanation* of this feat based on well-confirmed laws about the attributes and behavior of bats. In general, in order to obtain detailed understanding of a particular fact or of a law, it is necessary to have a good explanation of this fact or law. For instance, it is known that the leaves of certain types of trees change their color in the autumn. One can know this fact without understanding why it is the case. In order to understand why the leaves change color it is necessary to have an adequate explanation of why they do this.

In any area of scientific research we could record fact after fact, and still have no understanding of why these facts are true or how they are related to each other. In order to obtain an adequate understanding of particular facts and laws, we need to explain these particular facts and laws. In order to have a general understanding of a branch of science, we need a systematic, general, and hopefully unified, theory of this branch of science. To see this more clearly, suppose that a comprehensive theory of classical mechanics and gravitation had not been formulated until one hundred years after Newton's time. No doubt many more laws of terrestrial and solar system motions (like those of Galileo and Kepler) would have been discovered. But, to the very curious, these laws would have remained a puzzling collection of apparently unconnected facts. To understand them well requires a comprehensive theory containing some fundamental laws in terms of which the other, derivative laws can be explained.

Of course, in order to increase our understanding, we often must also increase our knowledge by discovering new laws which are used in the explanans of new explanations. In addition, as I indicated above, new surprising discoveries can often indicate gaps in our understanding. But knowing a fact or law is not the same as explaining this fact or law. Knowing a set of laws is not the same as having a systematic theory in which some laws are fundamental and others are explained, derivative laws. Increases in knowledge are closely related in a variety of ways to increases in understanding, and conversely. But knowledge and understanding are conceptually different, and we can certainly have much knowledge and still have little understanding. It is for this reason that basic research has the aim of increasing our knowledge and understanding, although a particular research project may be directed more towards one than the other. Similarly, in order to measure scientific progress, we must consider increases along the dimension of knowledge and also consider increases along the dimension of understanding. It would be an interesting and challenging project to investigate the various kinds of interdependencies and interactions between these two dimensions of scientific progress.

It is now time to be a little more specific in the discussion of understanding. Suppose that a particular branch of science is concerned with the entities in a certain domain. As this branch of science progresses, we will learn about various attributes of these entities and learn various laws about the behavior of various kinds of entities in this domain. With further progress, various kinds of entities in the domain will be systematically classified, and more general laws about the domain will be discovered. Some of these laws will be explained in terms of others, and eventually attempts will be made to construct a general, dynamic theory for this domain. Hopefully, in the long run, a unified, dynamic theory will be constructed and confirmed. If an acceptable unified theory is obtained, it will provide a high degree of understanding of its domain. Unified theories were extensively discussed in Chapter 6. The various ways in which they provide understanding should be fairly clear from the discussion in that chapter. However, it would be useful to emphasize here some of the more important aspects of the understanding provided by a unified theory.

In Chapter 6 I stated ten necessary conditions, U_1–U_{10}, which must be satisfied by a unified theory. Let T be such a theory. Then T has a set F of fundamental law-sentences, a set D of derivative law-sentences, and a set I of identities. The fundamental laws of T are left unexplained in T, but all of the derivative laws of T are explained in T. Thus, T provides understanding of its derivative laws in terms of its fundamental laws and its identities. T also provides a systematic classification of its ontology in terms of the basic and derivative predicates of its language. I believe that our understanding of a domain is enhanced by a systematic classification of that domain and its associated attributes. For example, consider the Periodic Table of the Elements. In order to develop this classification system, it was necessary to discover a large number of relationships among the properties of the elements. This involved considerable increase in the knowledge of chemistry. But, in addition, the systematic character of this knowledge gives it an informational content which seems to be far greater than that which would be expected from a similar, but unorganized, collection of facts. With the help of the Periodic Table, an experienced chemist can make an enormous number of useful predictions about elements and compounds. It is not at all obvious that these predictions are made by the use of DN explanations (although nowadays they could, in principle, be derived from the atomic-molecular theory). At any rate, it *appears* that a good, systematic classification system enhances our understanding of a domain and its associated attributes. It would be an interesting and challenging problem to develop a detailed analysis of the nature of this understanding. Perhaps it can be analyzed into the type of understanding provided by explanations.

A unified theory can provide understanding in additional, special ways not necessarily included among the general aspects mentioned above. Consider U_8. This condition requires that no basic predicate of the language of T occur in law-sentences of F only as an isolated correlate. If we were convinced that some theory must contain an isolated correlate, then this correlated basic predicate would have a mysterious role in the theory. Likewise, the correlation law-sentence in which it occurs as an isolated correlate would be an utterly mysterious fundamental law-sentence of the theory. The theory would contain a law which is not explained in the theory and which also would not play a significant role as a fundamental law of the theory. We would feel that there is a severe gap in our understanding. By excluding such isolated correlates, U_8 guarantees that we will not suffer this type of understanding gap with a unified theory. Strictly speaking, perhaps one should not say that a unified theory provides a special type of understanding in this way. It might be better to say that the theory avoids a certain type of lack of understanding. Similar remarks apply to several of the other conditions for unity. The following two cases are of particular interest.

By U_9 all true identities between predicates of the language of **T** are included in **I**. This guarantees that **T** will indicate the equivalence of two different descriptions of the same phenomena covered by **T**. If this condition is not satisfied, the phenomena of the theory would *appear* much more complex than they really are. We would discover various kinds of *apparent* parallelisms. These mysterious parallelisms would either remain unexplained, or else we would try to explain them with the help of spurious correlation laws. In the first case we would suffer a certain lack of understanding. In the second case we would have an apparent but false, understanding. However, because of U_9, a unified theory avoids both of these types of unsatisfactory circumstances. An adequate understanding of certain types of phenomena requires knowing certain identities. Because of U_9, having a unified theory will provide this understanding.

Now consider U_{10}. It is the most complex and subtle necessary condition for a unified theory. There are a variety of ways in which different kinds of theories might fail to satisfy this condition. To take a simple example, consider a theory with structured wholes. Suppose that this theory contains a basic property-predicate which, according to the theory, does not apply to any elements of Bas_1, but which does apply to some compound elements of the domain. The property denoted by this predicate would be an emergent property in this theory. Such properties may exist, but if they do, they would be very mysterious. If we were convinced of the existence of such an emergent property, we would also have to admit that there is a kind of gap in our understanding of the ontology of the theory. A unified theory avoids this kind of gap in our understanding.

It must be remembered that knowledge and understanding, although connected in many ways, are not the same. It is possible to have knowledge that we do not understand. It is also possible that some types of knowledge may indicate that certain types of understanding cannot be achieved. I have been describing some aspects of a kind of ideal understanding, and I have indicated ways in which a unified theory can help to provide this understanding. It would be possible to add further to this discussion of the understanding provided by a unified theory. However, the remarks above should be sufficient to demonstrate that a unified theory can provide considerable understanding. Now suppose that many different branches of science obtain such unified, dynamic theories. It is still possible to increase our understanding further by means of micro-reductions.

Let T_1, T_2 be unified theories such that T_1 is a theory with structured wholes and $Comp_1$ is identical with Dom_2. Let F_1, and F_2 be the sets of fundamental law-sentences of T_1, T_2, respectively. Suppose that we micro-reduce T_2 to T_1 according to the reduction conditions described in Chapter 5. In order to do this we must first discover and confirm suitable connecting sentences which are identities. Just doing this will no doubt require some increases in knowledge. Also, in order to accomplish the reduction, we must develop reductive explanations of the law-sentences in F_2 using as explanans the law-sentences in F_1 plus the connecting sentences.[1] Doing this results in understanding of the F_2 laws in terms of the F_1 laws. Indirectly, we also obtain a deeper understanding of the derivative laws of T_2. Consequently, we also learn of a broader scope of application of the F_1 laws. In addition, where we once had two sets of fundamental laws, we now have one set. Moreover, we unify two domains into one, and we simplify our ontology by identifying the things and attributes of T_2 with things and attributes of T_1. I believe that the ontological simplification and unification themselves provide an increase in understanding in addition to that provided by the reductive explanation of the laws of T_2. The ontological simplification and unification exhibit important systematic relationships between the two theories and their ontologies. As indicated above, although most understanding results from explanations, I believe that our understanding is also increased to some extent by the systematic organization of our knowledge. Whether or not one agrees with this, the microreduction certainly provides increase in understanding which results from its reductive explanations.

In this section I have tried only to indicate some very general features of scientific progress, and to exhibit an important role of the unification program in scientific progress. The unification program is based on successive microreductions of dynamic theories. In order for the unification program to progress, we must discover new knowledge in the form of connecting sentences. As the unification program progresses, our understanding of the world increases. Increases in knowledge and increases in understanding are essential requirements of scientific progress. The unification of dynamic theories thus plays an important role in scientific progress. Of course, some areas of research are concerned with developmental or evolutionary changes. Progress in this type of research must also be taken into consideration. In the next section I will briefly discuss how the unification program is related to such research.

B. DEVELOPMENT AND EVOLUTION

Throughout this book I have limited my analyses of theories, reduction, and unification to dynamic theories. However, scientists are also concerned with gaining knowledge and understanding of developmental and evolutionary processes. For example, we wish to know and understand the processes of development of a seed into a tree, and the processes of the growth of crystals. Similarly, we wish to know and understand the evolution of the chemical elements in the universe, and the evolution of the organic species on the earth. There are many other types of developmental and evolutionary changes studied in such fields as geology, psychology, and sociology. For the purposes of the present discussion, I will say that an *evolutionary* study is concerned with actual changes which take place in a particular past, present, or future time period in a particular thing or class of things. A *developmental* study is concerned with the types of changes which take place over time in certain kinds of things under certain conditions.

Let us first consider explanations of development. In Chapter 3, Section E, I discussed the kinds of laws which should be included in an adequate theory with structured wholes. If the elements of Bas_1 of such a theory can undergo various transformations or changes of state, then the theory should contain law-sentences describing these transformations and changes and the conditions under which they occur. The theory should also contain laws governing the rates of such changes under various conditions. In addition, the theory should contain laws stating conditions under which various kinds of compound elements do or do not exist. A dynamic theory with structured wholes is therefore a powerful theory. Suppose that a particular kind of thing or system of things in the domain of such a theory is not at equilibrium. Then, by using the laws just mentioned together with descriptions of boundary conditions (which may be changing with time), it will be possible to explain the various changes the thing or system undergoes. For instance, this is the general procedure which is used to explain such changes as radioactive decay processes and chemical reactions. Although some types of processes (such as the development of a hurricane and the growth of a tree) are very complex, they should be accounted for in the same general way when adequate dynamic theories are available. In principle, at least, explanations of developmental processes should be provided by the combined use of the laws of dynamic theories together with appropriate descriptions of external boundary conditions. Any specialized law of development, which applies under specified conditions, should be a derivative law in some appropriate dynamic theory. If such a developmental law describes development across domains, then the appropriate dynamic theory will be the theory resulting from the unification of the theories of these domains. Thus, the unification of certain theories may be a prerequisite for the explanation of certain types of developmental changes. The unification program may therefore be expected to play an important role in our understanding of certain types of developmental processes. This is another significant role of the unification program in scientific progress.

Studies of development are concerned with the types of changes which take place under certain conditions. Evolutionary studies are concerned with actual changes. We just saw that an explanation of development can be made in terms of appropriate laws of dynamic theories plus a description of the assumed external boundary conditions. In order to explain a particular past evolutionary change, one must, among other things, make use of descriptions of actual past conditions. For example, if one wants to explain the evolution of the element *iron* in the universe, or explain the evolution of the *elephant* on the earth, he must at least have some knowledge of the relevant past conditions. Explanations of evolutionary changes are therefore quite different from the types of explanations which are presented within a dynamic theory. They are also different from explanations of general types of developmental changes. The laws of a dynamic theory apply to certain kinds of entities which exist under certain kinds of conditions, but a dynamic theory does not explain or require the *actual existence* of any of the entities in its domain. It is only necessary that the entities in this domain can exist under certain specifiable conditions. A dynamic theory alone cannot provide explanations of the *actual existence* of particular entities or kinds of entities, although it might be said that a dynamic theory can explain their *possible existence* under certain conditions. Thus, for example, the laws of physics and chemistry cannot alone explain the evolution of the elephant. In particular, these laws cannot alone explain the evolution of an organism having the anatomical structure of elephants. Some may think that this shows that biology is irreducible to the laws of chemistry and physics. This would be a confusion about the nature of theories and reduction. Reduction is concerned with the explanation of the laws of one theory by the laws of another theory. A reductive explanation is not an evolutionary explanation of the actual existence of things. Indeed, the laws of a dynamic theory also cannot explain the actual existence of things. The general dynamic laws of biology cannot alone explain the evolution of the elephant for the simple reason that an evolutionary explanation must make use of actual particular boundary conditions. For further discussion of these and related issues see Causey (1969b), Hull (1974, pp. 139–141), and Ruse (1973, pp. 214–216).

Since evolutionary explanations make use of actual boundary conditions, these explanations do not fit directly into the unification program. Nevertheless, there is an important connection between the unification program and evolutionary explanations. In Chapter 6, Section A, I mentioned that some have used evolutionary considerations to try to show the advantages of microreductive explanations over macroreductive explanations. Suppose we have T_1 with Bas_1 and $Comp_1$, and T_2 with $Dom_2 = Comp_1$. Suppose that we know that at past time t_1 there existed elements of Bas_1 and no elements of $Comp_1$. Also we know that at past time $t_2 > t_1$, there also existed elements of $Comp_1$. For the reasons stated in Chapter 6, Section A, it would be extremely peculiar to macroreduce T_1 to T_2, but it would seem natural and reasonable to microreduce T_2 to T_1. Nevertheless, as I pointed out in that chapter, these considerations do not provide any proof that a microreduction is possible. It is logically possible that the things in $Comp_1 = Dom_2$ are emergent, and have emergent attributes, with respect to the fundamental laws of T_1. Evolutionary considerations provide some reason to attempt microreductions, but they provide no proof that success will be forthcoming.

On the other hand, it is interesting to consider the converse situation. Suppose that T_1 is a theory with structured wholes, and suppose that we have microreduced T_2 to T_1. T_1 has laws which state the conditions under which the kinds of elements in $Comp_1$ do or do not exist. Also, as we saw above, T_1 is therefore able to explain developmental changes which the elements in Dom_1 undergo under various conditions. Thus, T_1 is capable of specifying the kinds of conditions under which various elements in Dom_1 can change into other kinds of elements in Dom_1. In particular, T_1 is capable of specifying the kinds of conditions under which elements of Bas_1 can evolve into elements of $Comp_1 = Dom_2$. Hence, in order to explain the evolution of Dom_2 from Bas_1, we *should* use the laws of T_1 together with actual past conditions which existed between t_1 and t_2. More generally, we see that the microreduction of one branch of science to another is relevant to explanations of the evolution of the elements of the domain of one branch from those of the other. Still more generally, the unification program is relevant to the explanation of all sorts of evolutionary changes, within domains, across domains, from simpler to more complex things, and from more complex to simpler things. These observations are also important in another way. If the unification program is successful, then the general laws of dynamic theories *must* play an important role (together with actual boundary conditions) in the explanation of evolutionary changes. This is an interesting observation because there has been much controversy about the role of general laws in evolutionary explanations. For further discussion, with particular emphasis on biological evolution, see Hull (1974) and Ruse (1973).

A remark should be added about so-called "evolutionary theories". Sometimes this term seems to be used to refer to fairly specific historical hypotheses which are used in particular evolutionary explanations. An "evolutionary theory" in this sense is clearly not a dynamic theory, and it has no direct role in the unification program. Sometimes the term "evolutionary theory" is used to refer to a description of the general types of mechanisms (e.g., variations and natural selection) which are involved in an explanation of a certain general type of evolutionary change (e.g., evolution of biological species). This type of evolutionary theory is related in an obvious way to dynamic theories. The relevant dynamic theories must be able to explain why the hypothesized general mechanisms occur under the conditions under which they are supposed to occur. For example, the appropriate dynamic biological theory should be able to explain why variations occur in a large set of organisms of a given species. Also the appropriate dynamic biological theory should be able to characterize *fitness* and explain why natural selection occurs (in this case ecological factors are relevant).

Explanations of particular evolutionary changes, whether past, present, or future, require knowledge of actual conditions. There are special methodological problems connected with the acquisition and confirmation of the necessary factual data about these actual conditions. Moreover, explanations of evolutionary and developmental changes can be extremely complex. It would be out of place to discuss in this book the special problems connected with explanations of evolution and development. I do recognize that scientific progress involves in part the acquisition of knowledge and understanding of these processes. I have briefly indicated that the unification program plays an important, although somewhat indirect, role in our acquisition of this kind of knowledge and understanding.

C. PROBLEMS AND PROSPECTS

In Section A I characterized progress in basic research in terms of increases in knowledge and understanding. This book has been almost entirely concerned with dynamic theories. A dynamic theory grows as we discover more laws about the entities in the domain of this theory. Thus, as the theory grows, our knowledge of laws increases. In addition, when the theory is systematized into fundamental and derivative laws, our understanding increases by the explanations of the derivative in terms of the fundamental laws. If the theory is unified, our understanding increases further as a result of the increased nomological, explanatory, and ontological systematization. When two or more theories are unified by means of microreductions, or successive microreductions, our knowledge has increased further because of the discovery of the appropriate identity connecting sentences. In addition, our understanding is also increased as a result of the deeper reductive explanations and the resulting unification of the theories. Thus, it should now be clear that, in the case of dynamic theories, the discovery of new laws, theory construction, explanation, reduction, and unification all contribute to scientific progress.

In Section B I indicated some relationships between dynamic theories and the reduction program, on the one hand, and studies of evolutionary and developmental processes, on the other hand. It should be clear that increased knowledge and understanding of evolutionary and developmental processes are also components of scientific progress. Dynamic theories and the unification program can contribute to this kind of scientific progress.

Needless to say, describing the nature of scientific progress provides no guarantee that progress will be achieved. Yet, it seems clear to me that the history of science provides overwhelming evidence that much progress has taken place. Occasionally a theory which has been widely accepted will be overthrown, and we have a so-called scientific revolution. Such a theory may be abandoned (e.g., the phlogiston theory), or it may be reinterpreted and then subsumed under a more general and more accurate theory (e.g., the relationship between classical mechanics and special relativity theory). In both types of cases there is scientific progress. In at least the second kind of case one is also justified in claiming that some of the prerevolutionary knowledge and understanding is preserved through the revolution. No doubt there will be more scientific revolutions in the future. It is even conceivable that an extremely radical revolution could indicate severe limitations on further progress towards the unification of science. However, at the present time I see no compelling theoretical reasons to suppose that this will be the case. I am quite optimistic that science will continue to progress, including gradual unification, provided that research continues to receive adequate support.

It is sometimes thought that there may be theoretical limits on scientific progress which could be established *a priori*. I have yet to see a convincing argument which establishes a nontrivial, *a priori* limit. There may, of course, be empirical limitations of various kinds. There may be empirical limitations on laws, and on the kinds of theories, explanations, and reductions we can construct. There may certainly be practical limitations on our knowledge and explanation of particular events and conditions. It is appropriate to mention some of these kinds of possible limitations and the problems they can cause.

In the first place, it may be that some types of phenomena simply cannot be truly described by a general sentence which we would consider to be a law-sentence. In addition, we may only succeed in discovering relatively weak laws for certain types of phenomena. For instance, we may hope to find deterministic laws in a certain domain, and yet only discover probabilistic laws. I do not consider this to be a serious limitation, but it can lead to a vexing problem. If a probabilistic law is discovered in a certain domain, it may be difficult to decide whether it describes some intrinsically probabilistic phenomena, or whether the probabilistic character of the law is merely a reflection of our inability to control certain types of random disturbances. Although a question of this kind can be very difficult to settle, we should be prepared to accept the possibility that some laws are intrinsically probabilistic.[2]

Laws may be weak and somewhat limited in other ways. For instance, some phenomena may only be measurable on measurement scales which are weaker than either ratio or interval scales. Laws involving such quantities would be weaker than the kinds of physical laws with which we are now familiar. In such a case it is possible that further advances could lead to stronger measurement scales and to strengthened laws. However, in any given case it might be very difficult to decide whether or not such strengthening can be achieved.

Other kinds of possible empirical limitations can be imagined. For instance, we may simply not be able to construct a unified theory for a particular domain of phenomena. It may happen that the best theory we can construct simply does not satisfy all of the necessary conditions for a theory to be unified. It may be that the laws are such that no conceivable improvement of this theory would be a unified theory. Again, this would be an empirical limitation, and, as such, it is always possible (perhaps through a scientific revolution) that new discoveries could overcome it.

It is obvious that there are many possible empirical limitations which might be encountered by the program of unification through successive microreductions. In Chapter 7 I discussed what I believe to be the major obstacles which this program faces. I tried to show that, at the present time, none of these obstacles appears to be insurmountable. However, it is always possible that the growth of science may lead to the construction and confirmation of some theories which simply cannot be unified by means of microreductions. As has been previously indicated, I see no compelling reasons at the present time to expect this. However, it is possible, and if it happens, it would imply that there are certain types of limitations on our understanding of the world. But the unification program is a very large, long-range program, and we should certainly not allow set-backs to convince us quickly that such limitations really do exist. I expect that many of the most challenging and fascinating problems of future science will be problems connected with attempts to reduce one theory to another. This should be especially true if one of the theories uses probabilistic laws and the other does not, or if the two theories use very different types of measurement scales. Microreductions involve the study of structures, and we also face challenging problems in studies of the structures of nervous systems and in studies of social structures.

The possible empirical limitations described above all pertain directly to laws and dynamic theories. In Section B I indicated some connections between dynamic theories and evolutionary and developmental studies. Because of these connections, limitations on dynamic theories and their unification will impose certain limitations on evolutionary and developmental studies. In addition, our knowledge and understanding of certain particular evolutionary changes is likely to suffer from certain practical limitations. It is very difficult to obtain detailed and accurate knowledge about certain past conditions. As a consequence, our knowledge and understanding of certain evolutionary changes is likely, for practical reasons, to be limited to somewhat general and sketchy accounts. Applied science, in the form of prediction and control of future events and conditions, will also suffer certain practical limitations. For example, we may eventually learn many laws governing individual human behavior. Yet, outside of highly controlled, laboratory-type environments, it will remain very difficult to make detailed predictions about the future behavior of a particular individual. For various practical reasons we simply will not be able to know all of the relevant internal and external boundary conditions involved in the individual's behavior. Thus, scientific progress does not threaten the continuation of surprising, creative human achievements. On the contrary, since scientific progress provides us with both theoretical and practical powers, it should enhance our creative abilities and enable more people to develop their creative abilities. There is another point related to this one. As we gain understanding of the physical basis of human behavior, we will learn laws which govern the behavior of each of us to some extent. There surely will be general aspects which all people have in common. On the other hand, we will learn much about the structures of nervous systems, and how these systems exhibit genetic individualities and how they change with experiences. Although we will learn about aspects which we have in common, we will also learn much about the great extent to which we are individuals. No one's individuality should be threatened by the unification program.

I have been commenting on some of the broader implications of the unification program. In this context it is natural to ask how the success of this program would be related to the methodological principles upon which it is founded. Throughout this book I have made use of, and in some cases introduced and defended, certain general distinctions and principles. For instance, I have used certain principles of logic and mathematics, and I have defended and used certain principles concerning law-sentences, identities, explanations, reduction, and theory unification.

Ultimately these principles are used to define certain types of goals, e.g., theory reduction and theory unification. They are also applied in order to show what must be done in order to accomplish these goals. These principles are not defended by empirical evidence and they do not describe empirical facts (at least when they are interpreted strictly as methodological principles). Yet, if the unification program is carried out, then human behavior would be understood as governed by physical laws and as causally explainable (at least in the broad sense of causal explanation used in this book). All human behavior would then be reduced to certain types of complex physical events. It *might* be thought that these complex events are devoid of any significance other than their purely physical aspects. This would seem to imply that all human actions, including the use of language, are meaningless. Then the assertions of methodological principles would be meaningless. Thus it would appear that the successful use of my methodological principles would result in the destruction of their significance. Hence, it would appear that these principles are self-defeating.

The line of reasoning just presented is only my own very rough sketch of a kind of argument which might be developed. An argument along these lines has been presented by Grene (1971). A somewhat similar, but much more complex argument has been presented by Malcolm (1968). It would be superfluous for me to analyze these arguments at this point. My general position towards such arguments is a fairly clear consequence of the discussion in the last section of Chapter 7. It will be recalled that the discussion in that section emphasized the importance of conventions and social structures in social behavior.

Clearly, any adequate theory of human behavior must account for the difference between purely physical reflexes and behavior which is intentionalistic and conventionalized. Suppose that it turns out that all behavior is explainable in terms of physical laws. Then intentionalistic and conventional behavior is so explainable. But these explanations will also make use of complex physiological and social structural boundary conditions. The explanation of reflex movements will be much simpler and much different from explanations of intentional and conventional behavior. The intentional and conventional behavior will retain a complexity and significance which reflex movements do not have. Although intentional behavior will be causally explainable, it will still be intentional behavior. Although the fact that certain people use certain conventions will be causally explainable, these conventions will not thereby become meaningless. Intentional behavior which is guided in part by conventions must always retain its complex relationships with the behavior of other people within a social structure. If a person performs an action in accordance with certain conventions, then the fact that he follows these conventions will be involved in a causal explanation of the nature of his action. But these conventions and this action will still be significant to this person and also to others who are socially related to him. Moreover, these conventions and this action can still be assessed by concerned people according to other goals and standards they might have.

If the unification program succeeds, it does not follow that the theories constructed or the logical and methodological principles underlying these theories are meaningless. In principle, although probably not in practice, it would be possible to give a physicalistic explanation of why we came to use these principles and why we constructed these theories. But the theories and principles will remain significant. In fact, the success of the unification program would vindicate the methodological principles by showing that these principles have helped us to achieve the goal of a certain type of understanding of the world.

NOTES

[1] Since we are assuming here that $Dom_2 = Comp_1$, it is not likely that there will be many law-sentences in $\mathbf{F}_2$ which are n-equivalent to law-sentences in $\mathbf{F}_1$. Of course, if there are such n-equivalences, their discovery adds to our knowledge of the world. Such a discovery also affects our understanding in the manner indicated in Chapter 5, Section C.

[2] A useful comparison of different types of probabilistic laws can be found in Mackie (1974, Chapter 9).

BIBLIOGRAPHY

Achinstein, P.: 1971, *Law and Explanation*, Oxford University Press, Oxford.
Achinstein, P.: 1974, 'The Identity of Properties', *American Philosophical Quarterly* **11**, 257–275.
Adams, E. W.: 1959, 'The Foundations of Rigid.Body Mechanics and the Derivation of Its Laws from Those of Particle Mechanics', in L. Henkin *et al.* (eds.), *The Axiomatic Method*, North-Holland, Amsterdam, pp. 250–265.
Ager, T., J. Aronson, and R. Weingard: 1974, 'Are Bridge Laws Really Necessary?', *Noûs* **8**, 119–134.
Bearn, A. G.: 1971, 'Albinism', in P. Beeson and W. McDermott (eds.), *Textbook of Medicine*, Saunders, Philadelphia, p. 1698.
Bergmann, G.: 1966, *Philosophy of Science*, University of Wisconsin Press, Madison.
Block, N. J., and J. A. Fodor: 1972, 'What Psychological States Are Not', *The Philosophical Review* **81**, 159–181.
Brandt, R. and J. Kim: 1967, 'The Logic of the Identity Theory', *The Journal of Philosophy* **64**, 515–537.
Brodbeck, M.: 1958, 'Methodological Individualisms: Definition and Reduction', *Philosophy of Science* **25**, 1–22.
Bromberger, S.: 1966, 'Why-Questions', in R. Colodny (ed.), *Mind and Cosmos*, University of Pittsburgh Press, Pittsburgh, pp. 86–111.
Bunge, M.: 1967, *Scientific Research*, Vol. I, Springer-Verlag, New York.
Causey, R. L.: 1967, *Derived Measurement and the Foundations of Dimensional Analysis*, Measurement Theory and Mathematical Models Reports, Technical Report No. 5, University of Oregon, Eugene, Ore.
Causey, R. L.: 1968a, 'Complications in the Unification of Science', read at Philosophy of Science Association meeting, October, 1968, Pittsburgh, Pa.
Causey, R. L.: 1968b, 'The Importance of Being Surprised in Scientific Research', *Agricultural Science Review* **6**, 27–31.
Causey, R. L.: 1969a, 'Derived Measurement, Dimensions, and Dimensional Analysis', *Philosophy of Science* **36**, 252–270.
Causey, R. L.: 1969b, 'Polanyi on Structure and Reduction', *Synthese* **20**, 230–237.
Causey, R. L.: 1972a, 'Attribute-Identities in Microreductions', *The Journal of Philosophy* **69**, 407–422.
Causey, R. L.: 1972b, 'Uniform Microreductions', *Synthese* **25**, 176–218.
Causey, R. L.: 1976a, 'Identities and Reduction: A Reply', *Noûs* **10**, 333–337.
Causey, R. L.: 1976b, 'Unified Theories and Unified Science', in A. C. Michalos and R. S. Cohen (eds.), *PSA 1974* (Synthese Library), D. Reidel, Dordrecht, pp. 1–11.
Causey, R. L.: in press, 'Laws, Identities, and Reduction', in M. Przełecki *et al.* (eds.), *Formal Methods in the Methodology of Empirical Sciences: Proceedings of Conference for Formal Methods, Warsaw, June 17–21, 1974*, Ossolineum and D. Reidel, Wrocław and Dordrecht, forthcoming.
Dowben, R. M.: 1969, 'Composition and Structure of Membranes', in R. M. Dowben (ed.), *Biological Membranes*, Little, Brown & Co., Boston, pp. 1–38.
Enç, B.: 1976, 'Identity Statements and Microreductions', *The Journal of Philosophy* **73**, 285–306.

Feyerabend, P.: 1962, 'Explanation, Reduction, and Empiricism', in H. Feigl and G. Maxwell (eds.), *Minnesota Studies in the Philosophy of Science*, Vol. III, University of Minnesota Press, Minneapolis, Minn., pp. 28–97.

Feynman, R. P.: 1974, 'Structure of the Proton', *Science* **183**, 601–610.

Fodor, J. A.: 1974, 'Special Sciences (OR: The Disunity of Science as a Working Hypothesis)', *Synthese* **28**, 97–115.

Gendron, B.: 1971, 'On the Relation of Neurological and Psychological Theories: A Critique of the Hardware Thesis', in R. C. Buck and R. S. Cohen (eds.), *PSA 1970* (Synthese Library), D. Reidel, Dordrecht, pp. 483–495.

Grene, M.: 1971, 'Reducibility: Another Side Issue?', in M. Grene (ed.), *Interpretations of Life and Mind*, Routledge and Kegan Paul, London, pp. 14–37.

Hempel, C. G.: 1965, *Aspects of Scientific Explanation*, Free Press, New York.

Hempel, C. G.: 1966, *Philosophy of Natural Science*, Prentice-Hall, Englewood Cliffs, N.J.

Hempel, C. G.: 1969, 'Reduction: Ontological and Linguistic Facets', in S. Morgenbesser *et al.* (eds.), *Philosophy, Science, and Method*, St. Martin's, New York, pp. 179–199.

Howard, J. N.: 1964, 'John William Strutt, Third Baron Rayleigh', *Applied Optics* **3**, 1091–1101.

Hull, D.: 1974, *Philosophy of Biological Science*, Prentice-Hall, Englewood Cliffs, N.J.

Kalke, W.: 1969, 'What is Wrong with Fodor and Putnam's Functionalism?', *Noûs* **3**, 83–93.

Kemeny, J. G. and P. Oppenheim: 1956, 'On Reduction', *Philosophical Studies* **7**, 6–19.

Kim, J.: 1966, 'On the Psycho-Physical Identity Theory', *American Philosophical Quarterly* **3**, 227–235.

Kim, J.: 1969, 'Events and Their Descriptions: Some Considerations', in N. Rescher *et al.* (eds.), *Essays in Honor of Carl G. Hempel* (Synthese Library), D. Reidel, Dordrecht, pp. 198–215.

Krantz, D. H., R. D. Luce, P. Suppes, and A. Tversky: 1971, *Foundations of Measurement*, Vol. I, Academic Press, New York.

Kripke, S.: 1971, 'Identity and Necessity', in M. K. Munitz (ed.), *Identity and Individuation*, New York University Press, New York, pp. 135–164.

Lapedes, D. N. (ed.): 1971, *McGraw-Hill Encyclopedia of Science and Technology*, Vol. 3, McGraw-Hill, New York.

Levitt, I. M.: 1974, *Beyond the Known Universe*, Viking, New York.

Luce, R. D.: 1971, 'Similar Systems and Dimensionally Invariant Laws', *Philosophy of Science* **38**, 157–169.

Mackie, J. L.: 1974, *The Cement of the Universe*, Oxford University Press, Oxford.

Malcolm, N.: 1968, 'The Conceivability of Mechanism', *The Philosophical Review* **77**, 45–72.

Malinas, G. A.: 1973, 'Physical Properties', *Philosophia* **3**, 17–31.

Massey, G. J.: 1973, 'Reflections on the Unity of Science', *The Annals of the Japan Association for Philosophy of Science* **4**, 203–212.

Nagel, E.: 1961, *The Structure of Science*, Harcourt, Brace & World, New York.

Neurath, O. *et al.* (eds.): 1938, *International Encyclopedia of Unified Science*, Vols. I, II, University of Chicago Press, Chicago.

Nickles, T.: 1971, 'Covering Law Explanation', *Philosophy of Science* **38**, 542–561.

Oppenheim, P. and H. Putnam: 1958, 'Unity of Science as a Working Hypothesis', in H. Feigl *et al.* (eds.), *Minnesota Studies in the Philosophy of Science*, Vol. II, University of Minnesota Press, Minneapolis, pp. 3–36.

Partington, J. R.: 1953, *An Advanced Treatise on Physical Chemistry*, Vol. IV, Longmans, Green, London.

Pattee, H. H. (ed.): 1973, *Hierarchy Theory*, Braziller, New York.

Pelletier, F. J. (ed.): 1975, *Mass Terms*, *Synthese* **31**, 379–526.

Pitzer, K.: 1953, *Quantum Chemistry*, Prentice-Hall, Englewood Cliffs, N.J.

Platt, J. R.: 1961, 'Properties of Large Molecules That Go Beyond the Properties of Their Chemical Sub-Groups', *Journal of Theoretical Biology* **1**, 342–358.

Polanyi, M.: 1968, 'Life's Irreducible Structure', *Science* **160**, 1308–1312.

Putnam, H.: 1969, 'On Properties', in N. Rescher *et al.* (eds.), *Essays in Honor of Carl G. Hempel* (Synthese Library), D. Reidel, Dordrecht, pp. 235–254.

Quine, W. V. O.: 1960, *Word and Object*, M.I.T. Press, Cambridge, Mass.

Quine, W. V. O.: 1961, *From a Logical Point of View*, 2nd rev. ed., Harper & Row, New York.

Ruse, M.: 1973, *The Philosophy of Biology*, Hutchinson, London.

Schaffner, K.: 1967, 'Approaches to Reduction', *Philosophy of Science* **34**, 137–147.

Schaffner, K.: 1974, 'The Peripherality of Reductionism in the Development of Molecular Biology', *Journal of the History of Biology* **7**, 111–139.

Schlesinger, G.: 1963, *Method in the Physical Sciences*, Humanities Press, New York.

Sklar, L.: 1967, 'Types of Inter-Theoretic Reductions', *British Journal for the Philosophy of Science* **18**, 109–124.

Strawson, P. F.: 1963, *Individuals*, Anchor, Garden City, N.Y.

Suppes, P.: 1957, *Introduction to Logic*, Van Nostrand, New York.

Suppes, P. and J. L. Zinnes: 1963, 'Basic Measurement Theory', in R. D. Luce *et al.* (eds.), *Handbook of Mathematical Psychology*, Vol. I, John Wiley and Sons, New York, pp. 1–76.

Weiss, P. A.: 1970, 'Life, Order, and Understanding', *The Graduate Journal of the University of Texas* **8**, 1–157.

Weiss, P. A. (ed.): 1971, *Hierarchically Organized Systems in Theory and Practice*, Hafner Publ. Co., New York.

Whyte, L. L., A. G. Wilson, and D. Wilson (eds.): 1969, *Hierarchical Structures*, American Elsevier, New York.

Woodger, J. H.: 1952, *Biology and Language*, Cambridge University Press, Cambridge.

INDEX OF NAMES

INDEX OF SUBJECTS

emergent
  laws, *see* law(s), *ad hoc*
  property 65, 73, 76–78, 109, 166, 170
environmental conditions 66
ESS 53
event identities 145–151 *passim*
evolution and unification 109–110, 170–171
evolution, explanations of 169–172
evolutionary theories 2, 171
explanandum 18
explanans 18
explanation
  and understanding 163–164
  causal 25–27, 110, 112, 125–126
  D-N 23–27, 45–46, 94–95, 112
  D-S 27
  law-sentences subject to 35
  of actual and of possible existence
    169–170
  of behavior 152–159, 175–177
  of development 168–170
  of evolution 169–172
  vs. deduction 23–27, 94–95
explanatory derivation, *see* derivation,
    explanatory
explanatory-equivalence, e-eq 45–47, 92

F 45
free state 60, 67
fundamental law-sentences 27, 45, 67, 112
  derivation of 116–117

General Homogeneity Principle 63–64
genetics 139–141

hierarchical structures 138–142
Homogeneity Principle, General 63–64
homogeneous
  predicate 57
  sample of matter 51
  set 52, 57
  thing-predicate 52, 57, 63–65

I 45
identities 31–36
  and necessity 96–97
  confirmation of 98–99
  of attributes (attribute-) 32, 86–89, 93–94
  of events 145–151 *passim*
  of kinds (thing-) 32, 69–72, 80–83
  of quantities 40–42

quasi- 43–44
  synthetic 32–33, 36, 87, 95–96
identity *vs.* correlation 36, 80–83, 93–94
indirect reduction 141–142
isolated correlate 114, 118–119, 165–166

kinds 8–9, 59, 143–151 *passim*
  identities of 32, 69–72, 80–83
  of matter 51
  same 59–61, 63–64
knowledge 160–163, 164, 166–168

$\mathscr{L}$ 31, 37
law(s)
  ad hoc 67–68
  "bridge" 69, 93, 143
  general aspects of 13–16
  of theories with structured wholes 65–68
  probabilistic 15, 26, 42–43, 173–174
  quantitative 13–14, 26, 41–42
  unity of 106–107
  see also law-sentences
law-likeness, preservation of 29
law-sentences
  derivative 19, 27, 45
  explanatory independence of 116–117
  forms of 13–15
  fundamental 27, 45, 67, 112, 116–117
  interpretation of predicates in 28–30, 33,
    46, 79, 97–98
  nomological equivalence of, *see* NE,
    nomological-equivalence subject to
    causal explanation 35
  see also law(s)
logical truth 19–20
logic 7–8

macroreduction 110, 123–130
mass term 8–9, 51
measurement 10–13
  derived 10
  fundamental 10–12
  see also quantities
methodological individualism 152
microreduction(s)
  and evolution 109–110, 170–171
  conditions for 48–49, 58, 89–92
  connecting sentences of 54, 69–72, 75–78,
    80–89
  construction of 98–104
  explanations in 72–78, 84–89, 90–92

SYNTHESE LIBRARY

Monographs on Epistemology, Logic, Methodology,
Philosophy of Science, Sociology of Science and of Knowledge, and on the
Mathematical Methods of Social and Behavioral Sciences

Managing Editor:
JAAKKO HINTIKKA (Academy of Finland and Stanford University)

Editors:

ROBERT S. COHEN (Boston University)
DONALD DAVIDSON (University of Chicago)
GABRIËL NUCHELMANS (University of Leyden)
WESLEY C. SALMON (University of Arizona)

1. J. M. Bocheński, *A Precis of Mathematical Logic*. 1959, X + 100 pp.
2. P. L. Guiraud, *Problèmes et méthodes de la statistique linguistique*. 1960, VI + 146 pp.
3. Hans Freudenthal (ed.), *The Concept and the Role of the Model in Mathematics and Natural and Social Sciences, Proceedings of a Colloquium held at Utrecht, The Netherlands, January 1960*. 1961, VI + 194 pp.
4. Evert W. Beth, *Formal Methods. An Introduction to Symbolic Logic and the Study of Effective Operations in Arithmetic and Logic*. 1962, XIV + 170 pp.
5. B. H. Kazemier and D. Vuysje (eds.), *Logic and Language. Studies Dedicated to Professor Rudolf Carnap on the Occasion of His Seventieth Birthday*. 1962, VI + 256 pp.
6. Marx W. Wartofsky (ed.), *Proceedings of the Boston Colloquium for the Philosophy of Science, 1961-1962*, Boston Studies in the Philosophy of Science (ed. by Robert S. Cohen and Marx W. Wartofsky), Volume I. 1973, VIII + 212 pp.
7. A. A. Zinov'ev, *Philosophical Problems of Many-Valued Logic*. 1963, XIV + 155 pp.
8. Georges Gurvitch, *The Spectrum of Social Time*. 1964, XXVI + 152 pp.
9. Paul Lorenzen, *Formal Logic*. 1965, VIII + 123 pp.
10. Robert S. Cohen and Marx W. Wartofsky (eds.), *In Honor of Philipp Frank*, Boston Studies in the Philosophy of Science (ed. by Robert S. Cohen and Marx W. Wartofsky), Volume II. 1965, XXXIV + 475 pp.
11. Evert W. Beth, *Mathematical Thought. An Introduction to the Philosophy of Mathematics*. 1965, XII + 208 pp.
12. Evert W. Beth and Jean Piaget, *Mathematical Epistemology and Psychology*. 1966, XII + 326 pp.
13. Guido Küng, *Ontology and the Logistic Analysis of Language. An Enquiry into the Contemporary Views on Universals*. 1967, XI + 210 pp.
14. Robert S. Cohen and Marx W. Wartofsky (eds.), *Proceedings of the Boston Colloquium for the Philosophy of Science 1964-1966, in Memory of Norwood Russell Hanson*, Boston Studies in the Philosophy of Science (ed. by Robert S. Cohen and Marx W. Wartofsky), Volume III. 1967, XLIX + 489 pp.

15. C. D. Broad, *Induction, Probability, and Causation. Selected Papers.* 1968, XI + 296 pp.

16. Günther Patzig, *Aristotle's Theory of the Syllogism. A Logical-Philosophical Study of Book A of the Prior Analytics.* 1968, XVII + 215 pp.

17. Nicholas Rescher, *Topics in Philosophical Logic.* 1968, XIV + 347 pp.

18. Robert S. Cohen and Marx W. Wartofsky (eds.), *Proceedings of the Boston Colloquium for the Philosophy of Science 1966-1968*, Boston Studies in the Philosophy of Science (ed. by Robert S. Cohen and Marx W. Wartofsky), Volume IV. 1969, VIII + 537 pp.

19. Robert S. Cohen and Marx W. Wartofsky (eds.), *Proceedings of the Boston Colloquium for the Philosophy of Science 1966-1968*, Boston Studies in the Philosophy of Science (ed. by Robert S. Cohen and Marx W. Wartofsky), Volume V. 1969, VIII + 482 pp.

20. J.W. Davis, D. J. Hockney, and W. K. Wilson (eds.), *Philosophical Logic.* 1969, VIII + 277 pp.

21. D. Davidson and J. Hintikka (eds.), *Words and Objections: Essays on the Work of W. V. Quine.* 1969, VIII + 366 pp.

22. Patrick Suppes, *Studies in the Methodology and Foundations of Science. Selected Papers from 1911 to 1969*, XII + 473 pp.

23. Jaakko Hintikka, *Models for Modalities. Selected Essays.* 1969, IX + 220 pp.

24. Nicholas Rescher *et al.* (eds.), *Essays in Honor of Carl G. Hempel. A Tribute on the Occasion of His Sixty-Fifth Birthday.* 1969, VII + 272 pp.

25. P. V. Tavanec (ed.), *Problems of the Logic of Scientific Knowledge.* 1969, XII + 429 pp.

26. Marshall Swain (ed.), *Induction, Acceptance, and Rational Belief.* 1970, VII + 232 pp.

27. Robert S. Cohen and Raymond J. Seeger (eds.), *Ernst Mach: Physicist and Philosopher*, Boston Studies in the Philosophy of Science (ed. by Robert S. Cohen and Marx W. Wartofsky), Volume VI. 1970, VIII + 295 pp.

28. Jaakko Hintikka and Patrick Suppes, *Information and Inference.* 1970, X + 336 pp.

29. Karel Lambert, *Philosophical Problems in Logic. Some Recent Developments.* 1970, VII + 176 pp.

30. Rolf A. Eberle, *Nominalistic Systems.* 1970, IX + 217 pp.

31. Paul Weingartner and Gerhard Zecha (eds.), *Induction, Physics, and Ethics: Proceedings and Discussions of the 1968 Salzburg Colloquium in the Philosophy of Science.* 1970, X + 382 pp.

32. Evert W. Beth, *Aspects of Modern Logic.* 1970, XI + 176 pp.

33. Risto Hilpinen (ed.), *Deontic Logic: Introductory and Systematic Readings.* 1971, VII + 182 pp.

34. Jean-Louis Krivine, *Introduction to Axiomatic Set Theory.* 1971, VII + 98 pp.

35. Joseph D. Sneed, *The Logical Structure of Mathematical Physics.* 1971, XV + 311 pp.

36. Carl R. Kordig, *The Justification of Scientific Change.* 1971, XIV + 119 pp.

37. Milič Čapek, *Bergson and Modern Physics*, Boston Studies in the Philosophy of Science (ed. by Robert S. Cohen and Marx W. Wartofsky), Volume VII. 1971, XV + 414 pp.

38. Norwood Russell Hanson, *What I Do Not Believe, and Other Essays* (ed. by Stephen Toulmin and Harry Woolf), 1971, XII + 390 pp.

39. Roger C. Buck and Robert S. Cohen (eds.), *PSA 1970. In Memory of Rudolf Carnap*, Boston Studies in the Philosophy of Science (ed. by Robert S. Cohen and Marx W. Wartofsky), Volume VIII. 1971, LXVI + 615 pp. Also available as paperback.

40. Donald Davidson and Gilbert Harman (eds.), *Semantics of Natural Language.* 1972, X + 769 pp. Also available as paperback.

41. Yehoshua Bar-Hillel (ed.), *Pragmatics of Natural Languages.* 1971, VII + 231 pp.

42. Sören Stenlund, *Combinators, λ-Terms and Proof Theory.* 1972, 184 pp.

43. Martin Strauss, *Modern Physics and Its Philosophy. Selected Papers in the Logic, History, and Philosophy of Science.* 1972, X + 297 pp.

44. Mario Bunge, *Method, Model and Matter.* 1973, VII + 196 pp.

45. Mario Bunge, *Philosophy of Physics.* 1973, IX + 248 pp.

46. A. A. Zinov'ev, *Foundations of the Logical Theory of Scientific Knowledge (Complex Logic)*, Boston Studies in the Philosophy of Science (ed. by Robert S. Cohen and Marx W. Wartofsky), Volume IX. Revised and enlarged English edition with an appendix, by G. A. Smirnov, E. A. Sidorenka, A. M. Fedina, and L. A. Bobrova. 1973, XXII + 301 pp. Also available as paperback.

47. Ladislav Tondl, *Scientific Procedures*, Boston Studies in the Philosophy of Science (ed. by Robert S. Cohen and Marx W. Wartofsky), Volume X. 1973, XII + 268 pp. Also available as paperback.

48. Norwood Russell Hanson, *Constellations and Conjectures* (ed. by Willard C. Humphreys, Jr.). 1973, X + 282 pp.

49. K. J. J. Hintikka, J. M. E. Moravcsik, and P. Suppes (eds.), *Approaches to Natural Language. Proceedings of the 1970 Stanford Workshop on Grammar and Semantics.* 1973, VIII + 526 pp. Also available as paperback.

50. Mario Bunge (ed.), *Exact Philosophy – Problems, Tools, and Goals.* 1973, X + 214 pp.

51. Radu J. Bogdan and Ilkka Niiniluoto (eds.), *Logic, Language, and Probability. A Selection of Papers Contributed to Sections IV, VI, and XI of the Fourth International Congress for Logic, Methodology, and Philosophy of Science, Bucharest, September 1971.* 1973, X + 323 pp.

52. Glenn Pearce and Patrick Maynard (eds.), *Conceptual Chance.* 1973, XII + 282 pp.

53. Ilkka Niiniluoto and Raimo Tuomela, *Theoretical Concepts and Hypothetico-Inductive Inference.* 1973, VII + 264 pp.

54. Roland Fraïssé, *Course of Mathematical Logic* – Volume 1: *Relation and Logical Formula.* 1973, XVI + 186 pp. Also available as paperback.

55. Adolf Grünbaum, *Philosophical Problems of Space and Time.* Second, enlarged edition, Boston Studies in the Philosophy of Science (ed. by Robert S. Cohen and Marx W. Wartofsky), Volume XII. 1973, XXIII + 884 pp. Also available as paperback.

56. Patrick Suppes (ed.), *Space, Time, and Geometry.* 1973, XI + 424 pp.

57. Hans Kelsen, *Essays in Legal and Moral Philosophy*, selected and introduced by Ota Weinberger. 1973, XXVIII + 300 pp.

58. R. J. Seeger and Robert S. Cohen (eds.), *Philosophical Foundations of Science. Proceedings of an AAAS Program, 1969*, Boston Studies in the Philosophy of

Science (ed. by Robert S. Cohen and Marx W. Wartofsky), Volume XI. 1974, X + 545 pp. Also available as paperback.

59. Robert S. Cohen and Marx W. Wartofsky (eds.), *Logical and Epistemological Studies in Contemporary Physics*, Boston Studies in the Philosophy of Science (ed. by Robert S. Cohen and Marx W. Wartofsky), Volume XIII. 1973, VIII + 462 pp. Also available as paperback.

60. Robert S. Cohen and Marx W. Wartofsky (eds.), *Methodological and Historical Essays in the Natural and Social Sciences. Proceedings of the Boston Colloquium for the Philosophy of Science, 1969-1972,* Boston Studies in the Philosophy of Science (ed. by Robert S. Cohen and Marx W. Wartofsky), Volume XIV. 1974, VIII + 405 pp. Also available as paperback.

61. Robert S. Cohen, J. J. Stachel and Marx W. Wartofsky (eds.), *For Dirk Struik. Scientific, Historical and Political Essays in Honor of Dirk J. Struik*, Boston Studies in the Philosophy of Science (ed. by Robert S. Cohen and Marx W. Wartofsky), Volume XV. 1974, XXVII + 652 pp. Also available as paperback.

62. Kazimierz Ajdukiewicz, *Pragmatic Logic,* transl. from the Polish by Olgierd Wojtasiewicz. (1974, XV + 460 pp.

63. Sören Stenlund (ed.), *Logical Theory and Semantic Analysis. Essays Dedicated to Stig Kanger on His Fiftieth Birthday.* 1974, V + 217 pp.

64. Kenneth F. Schaffner and Robert S. Cohen (eds.), *Proceedings of the 1972 Biennial Meeting, Philosophy of Science Association,* Boston Studies in the Philosophy of Science (ed. by Robert S. Cohen and Marx W. Wartofsky), Volume XX. 1974, IX + 444 pp. Also available as paperback.

65. Henry E. Kyburg, Jr., *The Logical Foundations of Statistical Inference.* 1974, IX + 421 pp.

66. Marjorie Grene, *The Understanding of Nature: Essays in the Philosophy of Biology,* Boston Studies in the Philosophy of Science (ed. by Robert S. Cohen and Marx W. Wartofsky), Volume XXIII. 1974, XII + 360 pp. Also available as paperback.

67. Jan M. Broekman, *Structuralism: Moscow, Prague, Paris.* 1974, IX + 117 pp.

68. Norman Geschwind, *Selected Papers on Language and the Brain,* Boston Studies in the Philosophy of Science (ed. by Robert S. Cohen and Marx W. Wartofsky), Volume XVI. 1974, XII + 549 pp. Also available as paperback.

69. Roland Fraïssé, *Course of Mathematical Logic* – Volume 2: *Model Theory.* 1974, XIX + 192 pp.

70. Andrzej Grzegorczyk, *An Outline of Mathematical Logic. Fundamental Results and Notions Explained with All Details.* 1974, X + 596 pp.

71. Franz von Kutschera, *Philosophy of Language.* 1975, VII + 305 pp.

72. Juha Manninen and Raimo Tuomela (eds.), *Essays on Explanation and Understanding. Studies in the Foundations of Humanities and Social Sciences.* 1976, VII + 440 pp.

73. Jaakko Hintikka (ed.), *Rudolf Carnap, Logical Empiricist. Materials and Perspectives.* 1975, LXVIII + 400 pp.

74. Milič Čapek (ed.), *The Concepts of Space and Time. Their Structure and Their Development,* Boston Studies in the Philosophy of Science (ed. by Robert S. Cohen and Marx W. Wartofsky), Volume XXII. 1976, LVI + 570 pp. Also available as paperback.

75. Jaakko Hintikka and Unto Remes, *The Method of Analysis. Its Geometrical Origin and Its General Significance*, Boston Studies in the Philosophy of Science (ed. by Robert S. Cohen and Marx W. Wartofsky), Volume XXV. 1974, XVIII + 144 pp. Also available as paperback.

76. John Emery Murdoch and Edith Dudley Sylla, *The Cultural Context of Medieval Learning. Proceedings of the First International Colloquium on Philosophy, Science, and Theology in the Middle Ages – September 1973*, Boston Studies in the Philosophy of Science (ed. by Robert S. Cohen and Marx W. Wartofsky), Volume XXVI. 1975, X + 566 pp. Also available as paperback.

77. Stefan Amsterdamski, *Between Experience and Metaphysics. Philosophical Problems of the Evolution of Science*, Boston Studies in the Philosophy of Science (ed. by Robert S. Cohen and Marx W. Wartofsky), Volume XXXV. 1975, XVIII + 193 pp. Also available as paperback.

78. Patrick Suppes (ed.), *Logic and Probability in Quantum Mechanics.* 1976, XV + 541 pp.

79. H. von Helmholtz, *Epistemological Writings.* (A New Selection Based upon the 1921 Volume edited by Paul Hertz and Moritz Schlick, Newly Translated and Edited by R. S. Cohen and Y. Elkana), Boston Studies in the Philosophy of Science, Volume XXXVII. 1977 (forthcoming).

80. Joseph Agassi, *Science in Flux*, Boston Studies in the Philosophy of Science (ed. by Robert S. Cohen and Marx W. Wartofsky), Volume XXVIII. 1975, XXVI + 553 pp. Also available as paperback.

81. Sandra G. Harding (ed.), *Can Theories Be Refuted? Essays on the Duhem-Quine Thesis.* 1976, XXI + 318 pp. Also available as paperback.

82. Stefan Nowak, *Methodology of Sociological Research: General Problems.* 1977, XVIII + 504 pp. (forthcoming).

83. Jean Piaget, Jean-Blaise Grize, Alina Szeminska, and Vinh Bang, *Epistemology and Psychology of Functions.* 1977 (forthcoming).

84. Marjorie Grene and Everett Mendelsohn (eds.), *Topics in the Philosophy of Biology*, Boston Studies in the Philosophy of Science (ed. by Robert S. Cohen and Marx W. Wartofsky), Volume XXVII. 1976, XIII + 454 pp. Also available as paperback.

85. E. Fischbein, *The Intuitive Sources of Probabilistic Thinking in Children.* 1975, XIII + 204 pp.

86. Ernest W. Adams, *The Logic of Conditionals. An Application of Probability to Deductive Logic.* 1975, XIII + 156 pp.

87. Marian Przełęcki and Ryszard Wójcicki (eds.), *Twenty-Five Years of Logical Methodology in Poland.* 1977, VIII + 803 pp. (forthcoming).

88. J. Topolski, *The Methodology of History.* 1976, X + 673 pp. (forthcoming).

89. A. Kasher (ed.), *Language in Focus: Foundations, Methods and Systems. Essays Dedicated to Yehoshua Bar-Hillel*, Boston Studies in the Philosophy of Science (ed. by Robert S. Cohen and Marx W. Wartofsky), Volume XLIII. 1976, XXVIII + 679 pp. Also available as paperback.

90. Jaakko Hintikka, *The Intentions of Intentionality and Other New Models for Modalities.* 1975, XVIII + 262 pp. Also available as paperback.

91. Wolfgang Stegmüller, *Collected Papers on Epistemology, Philosophy of Science and History of Philosophy*, 2 Volumes, 1977 (forthcoming).

92. Dov M. Gabbay, *Investigations in Modal and Tense Logics with Applications to Problems in Philosophy and Linguistics*. 1976, XI + 306 pp.
93. Radu J. Bogdan, *Local Induction*. 1976, XIV + 340 pp.
94. Stefan Nowak, *Understanding and Prediction: Essays in the Methodology of Social and Behavioral Theories*. 1976, XIX + 482 pp.
95. Peter Mittelstaedt, *Philosophical Problems of Modern Physics*, Boston Studies in the Philosophy of Science (ed. by Robert S. Cohen and Marx W. Wartofsky), Volume XVIII. 1976, X + 211 pp. Also available as paperback.
96. Gerald Holton and William Blanpied (eds.), *Science and Its Public: The Changing Relationship*, Boston Studies in the Philosophy of Science (ed. by Robert S. Cohen and Marx W. Wartofsky), Volume XXXIII. 1976, XXV + 289 pp. Also available as paperback.
97. Myles Brand and Douglas Walton (eds.), *Action Theory. Proceedings of the Winnipeg Conference on Human Action, Held at Winnipeg, Manitoba, Canada, 9-11 May 1975*. 1976, VI + 345 pp.
98. Risto Hilpinen, *Knowledge and Rational Belief*. 1978 (forthcoming).
99. R. S. Cohen, P. K. Feyerabend, and M. W. Wartofsky (eds.), *Essays in Memory of Imre Lakatos*, Boston Studies in the Philosophy of Science (ed. by Robert S. Cohen and Marx W. Wartofsky), Volume XXXIX. 1976, XI + 762 pp. Also available as paperback.
100. R. S. Cohen and J. Stachel (eds.), *Leon Rosenfeld, Selected Papers*. Boston Studies in the Philosophy of Science (ed. by Robert S. Cohen and Marx W. Wartofsky), Volume XXI. 1977 (forthcoming).
101. R. S. Cohen, C. A. Hooker, A. C. Michalos, and J. W. van Evra (eds.), *PSA 1974: Proceedings of the 1974 Biennial Meeting of the Philosophy of Science Association*, Boston Studies in the Philosophy of Science (ed. by Robert S. Cohen and Marx W. Wartofsky), Volume XXXII. 1976, XIII + 734 pp. Also available as paperback.
102. Yehuda Fried and Joseph Agassi, *Paranoia: A Study in Diagnosis*, Boston Studies in the Philosophy of Science (ed. by Robert S. Cohen and Marx W. Wartofsky), Volume L. 1976, XV + 212 pp. Also available as paperback.
103. Marian Przełęcki, Klemens Szaniawski, and Ryszard Wójcicki (eds.), *Formal Methods in the Methodology of Empirical Sciences*. 1976, 455 pp.
104. John M. Vickers, *Belief and Probability*. 1976, VIII + 202 pp.
105. Kurt H. Wolff, *Surrender and Catch: Experience and Inquiry Today*, Boston Studies in the Philosophy of Science (ed. by Robert S. Cohen and Marx W. Wartofsky), Volume LI. 1976, XII + 410 pp. Also available as paperback.
106. Karel Kosík, *Dialectics of the Concrete*, Boston Studies in the Philosophy of Science (ed. by Robert S. Cohen and Marx W. Wartofsky), Volume LII. 1976, VIII + 158 pp. Also available as paperback.
107. Nelson Goodman, *The Structure of Appearance*, Boston Studies in the Philosophy of Science (ed. by Robert S. Cohen and Marx W. Wartofsky), Volume L. 1977 (forthcoming).
108. Jerzy Giedymin (ed.), *Kazimierz Ajdukiewicz: Scientific World-Perspective and Other Essays, 1930-1963*. 1977 (forthcoming).
109. Robert L. Causey, *Unity of Science*. 1977, VIII + 180 pp. + indices (forthcoming).
110. Richard Grandy, *Advanced Logic for Applications*. 1977 (forthcoming).

111. Robert P. McArthur, *Tense Logic*. 1976, VII + 84 pp.
112. Lars Lindahl, *Position and Change: A Study in Law and Logic*. 1977, IX + 299 pp.
113. Raimo Tuomela, *Dispositions*. 1977 (forthcoming).
114. Herbert A. Simon, *Models of Discovery and Other Topics in the Methods of Science*, Boston Studies in the Philosophy of Science (ed. by Robert S. Cohen and Marx W. Wartofsky), Volume LIV. 1977 (forthcoming).
115. Roger D. Rosenkrantz, *Inference, Method and Decision*. 1977 (forthcoming).
116. Raimo Tuomela, *Human Action and Its Explanation. A Study on the Philosophical Foundations of Psychology*. 1977 (forthcoming).
117. Morris Lazerowitz, *The Language of Philosophy*, Boston Studies in the Philosophy of Science (ed. by Robert S. Cohen and Marx W. Wartofsky), Volume LV. 1977 (forthcoming).
118. Tran Duc Thao, *Origins of Language and Consciousness*, Boston Studies in the Philosophy of Science (ed. by Robert S. Cohen and Marx. W. Wartofsky), Volume LVI. 1977 (forthcoming).